CAMBRIDGE COUNTY GEOGRAPHIES

SCOTLAND

General Editor: W. MURISON, M.A.

T0352325

KIRKCUDBRIGHTSHIRE

AND

WIGTOWNSHIRE

Cambridge County Geographies

KIRKCUDBRIGHTSHIRE
AND
WIGTOWNSHIRE

by

WILLIAM LEARMONTH, F.R.P.S., F.B.S.E.

Girthon Public School, Gatehouse-of-Fleet

With Maps, Diagrams, and Illustrations

CAMBRIDGE
AT THE UNIVERSITY PRESS
1920

CAMBRIDGE UNIVERSITY PRESS

Cambridge, New York, Melbourne, Madrid, Cape Town,
Singapore, São Paulo, Delhi, Mexico City

Cambridge University Press
The Edinburgh Building, Cambridge CB2 8RU, UK

Published in the United States of America by Cambridge University Press, New York

www.cambridge.org
Information on this title: www.cambridge.org/9781107675995

First published 1920
First paperback edition 2013

A catalogue record for this publication is available from the British Library

ISBN 978-1-107-67599-5 Paperback

CONTENTS

CONTENTS

ILLUSTRATIONS

ILLUSTRATIONS

MAPS

The illustrations on pp. 10, 11, 17, 22, 25, 26, 27, 35, 45, 47, 50, 51, 52, 54, 60, 64, 78, 95, 102, 104, 106, 107, 109, 111, 112, 113, 114, 118, 119, 120, 121, 122, 132, 133, 138, 140, 141, 146 are reproduced from photographs by the Rev. C. H. Dick ; those on pp. 6, 56, 84, 91, 103, 116 from photographs supplied by Messrs Valentine & Son, Ltd. ; that on p. 5 is reproduced by kind permission of *The Courier and Herald*, Dumfries ; that on p. 134 from a print kindly supplied by T. Fraser, Esq. ; those on pp. 143 and 145 from photographs supplied by The Wigtownshire Creamery Company.

1. County and Shire. The Origin of Galloway, Kirkcudbright, Wigtown

The word *shire* is of Old English origin and meant office, charge, administration. The Norman Conquest introduced the word *county*—through French from the Latin *comitatus*, which in mediaeval documents designates the shire. *County* is the district ruled by a count, the king's *comes*, the equivalent of the older English term *earl*. This system of local administration entered Scotland as part of the Anglo-Norman influence that strongly affected our country after the year 1100.

Galloway to-day, the Grey Galloway of literature, comprises the counties of Wigtown and Kirkcudbright. From east to west it extends from the " Brig en' o' Dumfries to the Braes o' Glenapp," or almost to the Braes, the western boundary of Wigtownshire at this part being, in point of fact, the Galloway Burn. In ancient times the Province of Galloway is said to have extended also over parts of the adjacent counties. But for hundreds of years the name has been identified solely with the " Stewartry " of Kirkcudbright and the " Shire " of Wigtown.

The origin of these terms dates back to 1369, when Archibald the Grim, third Earl of Douglas, received the lordship of Galloway, and the whole of the Crown lands

A   1

between the Nith and the Cree.   Archibald appointed
a steward to collect his revenues and administer justice,
whence the name *Stewartry*.   In the following year he
obtained Wigtownshire by purchase from the Earl of
Wigtown.   This district continued to be administered
by the King's Sheriff, and has been known ever since
as the *Shire*.   According to Skene in his *Celtic Scotland*
the word *Galloway* is formed by the combination of the
two words *Gall*, a stranger, and *Gaidhel*, the Gaels.
Gallgaidhel was the name given to the mixed Norse and
Gaels in the Hebrides, Man, Kintyre and Galloway.
To the last district the designation came latterly to be
restricted.   The word *Gallgaidhel* appears in Welsh as
Gallwyddel (where *dd* is pronounced as *th*), whence arose
the forms *Gallwitheia*, *Gallwitha*, *Gallovidia*, and *Galloway*.

The name *Kirkcudbright* means Cuthbert's Kirk.   The
same meaning belongs to the Gaelic term *Kilcudbrit*.
Bede records a visit of St Cuthbert to the Niduari, the
men of the region of the Nith.

Wigtown means bay-town, the first syllable being
from the Scandinavian *vik*, a bay, a creek.

Wigtownshire and Kirkcudbrightshire were two of the
three counties on whose boundaries, county and parish,
no change was made by the Commissioners under the
Act of 1889.

2. General Characteristics

Geographically, Galloway may be viewed as falling
into three divisions—Upper Galloway, the hilly northern

portions of both counties ; Lower Galloway, the lower
and more open southern sections of both divisions east
of Luce Bay ; and the Rhinns, the double peninsula to
the south-west of Luce Bay and Loch Ryan. There is
a quaint Latin description of Galloway written by John
MacLellan in 1665 for Blaeu's atlas, which may be thus
translated : " The whole region is very healthy in
climate and soil ; it rarely ascends into mountains, but
rises in many hills. Galloway as a whole recalls the
figure of an elephant ; the Rhinns form the head, the
Mull the proboscis ; the headlands jutting into the sea
the feet ; the mountains above-named the shoulders ;
rocks and moors the spine ; the remainder of the district
the rest of the body."

With a coastline of over 170 miles, its fishing is of
comparatively little importance ; its harbours are few,
and the bulk of its commerce is railway-borne ; while
the absence of coal and iron has reduced its manufactur-
ing industries to a minimum. Its wealth lies in its
agriculture. In the uplands sheep-rearing, in the low-
lands dairying and mixed farming give Kirkcudbrightshire and Wigtownshire a high place among the counties
of Scotland. Certain districts—Twynholm, Kirkcudbright, Borgue, Glenluce—pay much attention to bee-
keeping, and there the honey is not excelled by any
produced elsewhere in the British Isles.

Galloway offers many a bid for the outside world.
Its manifold beauty of storm-scarped mountain and
quiet loch ; its rivers, here brawling torrents, there
smooth-flowing streams ; its long seaboard of frowning

cliff relieved by sandy beach, woo the lover of nature with charms that will not be gainsaid. In many a fort and cairn, in many a mote and sculptured stone, the antiquary finds exposed the unwritten record of the storied past. Its once stately abbeys, whose ruins to-day invite the ecclesiologist, were centres of missionary effort which kept alive the torch of religion in the dark ages. Monuments on its whaup-haunted moors and tombstones in its " Auld Kirkyards " tell of the dour westland whigs and their part in Scotland's fight for religious freedom. Broken castle walls speak of long generations of " Neighbour Weir," as the feuds of the petty chiefs were oddly called. The charm of letters is not wanting. In Gatehouse-of-Fleet Burns is said to have committed to paper the flaming battle-ode which had sung itself into his soul to the accompaniment of a thunderstorm on the moor. Crockett's novels derive from the soil which gave him birth, and will long hold their place as typical of Galloway—its scenery, its people, and their homely hospitable ways. But a greater than Crockett has been here ; Scott found subjects in Galloway for *Guy Mannering, Old Mortality,* and *The Bride of Lammermoor,* while Jeanie Deans, the heroine of *The Heart of Midlothian,* had her prototype in Helen Walker, the daughter of a small farmer in the parish of Irongray.

Galloway affords ample scope for the labours of the geologist and the botanist, and presents varied and unnumbered subjects for the canvas of the artist. Add to this the possession of a climate so mild and equable that " the tulip tree flourishes and flowers at St Mary's

Glenluce Abbey

Isle, and the arbutus bears fruit at Kirkdale " ; and it will be readily conceded that " there is no district in Scotland better worth knowing."

Dalry, New Galloway and Carsphairn among the hills,

Portpatrick, looking South

and Stranraer, Portpatrick and Rockcliffe by the sea are but a few of the holiday haunts for which the district is noted.

3. Size. Shape. Boundaries

The longest straight line which can be measured across Kirkcudbrightshire is from Arbigland to a point on the Cree where the river separates the county from

Ayrshire. This runs, roughly speaking, from south-east to north-west, and is 44½ miles long. From a point a mile and a half north of Maxwelltown due west to the same river the length is 40¼ miles, while a line due north from the Ross promontory to the Dumfriesshire boundary is 37 miles. The area of the county is 575,832 acres. Among the counties in Scotland it is ninth in size. It is 1⅘th times the size of Wigtownshire, while it is only $\frac{7}{10}$ths that of Ayrshire and $\frac{8}{10}$ths the size of Dumfriesshire. The shape of Kirkcudbrightshire is very irregular, but is approximately trapezoidal.

On the south Kirkcudbrightshire is bounded by the Solway Firth and Wigtown Bay. On the west the Cree separates it first from Wigtownshire, and then from Ayrshire as far as Loch Moan. The dividing line runs east by the Merrick, to be continued by Eglin Lane, Loch Enoch, and Gala Lane to Loch Doon. For about half its length this loch is the county boundary. Northwards and then eastwards as far as Blacklarg, where the Stewartry meets Dumfriesshire, and southwards to the parish of Irongray, the boundary is mostly artificial. The Cairn Water, sweeping round Irongray and Terregles to its confluence with the Nith, about a mile and a quarter north of Dumfries, forms once more a natural boundary, which is continued by the Nith to the sea.

From Grange of Cree westwards through Stranraer to the North Channel, the extreme length of Wigtownshire is 30 miles ; its breadth from Burrow Head to the Ayrshire boundary is 31 miles. Thus were it not for Luce Bay and Loch Ryan the outline of the county would be approximately a rhombus. Exclusive of water, its area is 311,984 acres, and this is seventeenth in size among the counties of Scotland. It is barely half the size of Ayrshire, while it is rather more than half that of the Stewartry. From Carrick Mill Burn, where the three counties of Ayr, Kirkcudbright and Wigtown meet, the River Cree, with its estuary broadening into Wigtown Bay, forms the eastern boundary of Wigtown-shire. On the north the boundary runs eastward from Galloway Burn to the Main Water of Luce. Then bending for a short distance to the north, it cuts the Cross Water of Luce and, sweeping round Benbrake Hill, passes along Pulganny Burn, Loch Maberry and Loch Dornal to Carrick Mill Burn, where it meets the Cree.

Elsewhere, Wigtownshire is washed by the sea.

4. Surface and General Features

If a straight line be drawn from the middle of the parish of Irongray to the middle of Anwoth, it will be found that the land to the south-east is, on the whole, lowland in character ; that to the north-west is high-land. Yet the former is lowland only by contrast. An elevated tract of ground stretches from Criffel (1867) north-west by the Cuil Hill (1377) and the Long Fell to the Lotus Hill (1050). West of this the land gradually decreases in height towards the plain of the Urr. Again it rises in a ridge of rugged hills strewn with boulders to culminate in the Screel (1120) and Bengairn (1250).

Twynholm may be looked on as an elevated plain, so high in general does the surface of the parish lie. In the northern part of the parish is Fuffock Hill (1000), and Ben Gray (1200), which slopes down to Loch Whinyeon. The rest of this southern division, from the valley of the Dee eastwards to Terregles, while here and there hilly, is marked as a rule by an unbroken surface. Girthon parish is mostly bleak, heathery upland, consisting of broad irregular masses of hills intersected by water courses. These reach their greatest altitude in Craigronald (1684).

The highland district of Kirkcudbrightshire forms part of the wide table-land extending from St Abbs Head to Portpatrick, and known as the southern uplands of Scotland. It rises into a cluster of mountains with smooth tops, and sides scarped with precipices or deeply cut into with glens, presenting everywhere scenes of naked and rugged grandeur. Here are few trees ; here is but little trace of man. Nature is everywhere stern ; no cultivation is possible, and the region forms one vast sheep-walk, clad with heath and moss, relieved by stretches of eagerly sought-for grass. The interior of Kirkmabreck is a congeries of hills, of which the highest is Cairnsmore of Fleet (2331), partly in Minnigaff. The surface of this parish is everywhere mountainous. From south to north are Cairnsmore of Fleet, Larg Hill (2216), Lamachan (2349), Benyellary (2360), Merrick (2764) and Kirriereoch (2562). Merrick is the loftiest summit south of the Grampains. " Ony shauchle," was Crockett's inscription in one of his novels presented to a shepherd, " Ony shauchle can write a

Rocks near Loch Enoch

book, but it takes a man to herd the Merrick." An
undulating line connects the tops of these hills in one

wide sweep of tableland.   Sir Archibald Geikie describes
the surface of this parish as " one wild expanse of
mountain and moorland roughened with thousands of
heaps of glacial detritus, and dotted with lakes enclosed
within these rubbish mounds."   The heathy summits
of the Rhinns of Kells command a magnificent view.
From Little Millyea (1898) the range runs N.N.W.

Loch Enoch and Merrick

through Meikle Millyea (2446), Millfire (2350), and Corscrine (2668) to Coran of Portmark (2042).   In Carsphairn, with its lofty hills green almost to the top,
rearing every year no fewer than 30,000 Cheviot and
blackfaced sheep, the highest is Cairnsmore, the third
peak in the well-known lines :

"Cairnsmore of Fleet, and Cairnsmore of Dee,
  Cairnsmore of Carsphairn, the highest o' the three."

The second of these is in Kells, and is 1616 feet high.

No county in Scotland rises so little in the aggregate above the level of the sea as Wigtownshire, yet as a whole it is undulating and hillocky. The higher grounds in general are arranged in no regular order, the most important of them occurring as solitary eminences. The peninsula which stretches from Corsewall Point to Mull of Galloway is known as the Rhinns (Celtic, *rinn*, a point, with English plural); the bluntly triangular peninsula terminating in Burrow Head is called the Machers (Celtic, *mahair*, a plain, with English plural); the rest of the county stretching from the Cree to Loch Ryan and including a large part of the parishes of Penninghame, Kirkcowan, Mochrum, Inch, Old Luce and New Luce, bears the name of the Moors. Wild, and for the most part uncultivated, the moors, to which the name is due, are the chief feature in its scenery. They are well stocked with game, but except for sheep-farming are of no value industrially. It is possible to travel from Glenhapple Moor, near the Cree, westwards through Urrall, Dirneark, Airieglasson, Laggangairn, Glenkitten, Dalnyap, Mark and Laight to Loch Ryan, in some cases over " flows " (as the peat mosses are called) from eight to ten miles long, without crossing a single ploughed field. Where there is cultivation it is confined almost entirely to narrow strips along the courses of some of the streams. In the north of the Moors are the highest hills of the county—Midmoile (844), Craigairie and Benbrake (each 1000).

Lying south of the Moors, from which it is separated by no well-defined boundary, is the peninsula called the Machers. It comprises the parishes of Wigtown, Kirkinner, Sorbie, Whithorn, Glasserton, most of Mochrum, and parts of Old Luce. The surface as a rule is low and flat, but the general flatness is relieved here and there by gently sloping ridges running with a fairly uniform trend from north-east to south-west, and rising as a whole towards the south-west. The highest elevations, all near the coast on that side, are Fell of Carleton (475), Fell of Barhullion (450), East Bar (450), Bennan Hill (500), Mochrum Fell (646), Doon of May (457) ; and, on account of its position and configuration more conspicuous than hills which are higher, Knock of Luce (513). The district is well watered. Of the numerous streams it is noteworthy that the larger, *e.g.* the Ket and the Drummullin Burn, run transversely to the general trend of ridges and hollows, while the smaller flow between the ridges. A feature of the district is the manner in which the boulder clay is scattered over the ground. The large, nearly oblong, smooth ridges of this deposit, known locally as " drums," are invariably cultivated, and the contrast between them and the surrounding lower and uncultivated ground is very striking.

Connected with the rest of the county by an isthmus six miles broad at its narrowest part, the double peninsula of the Rhinns measures 28¼ miles from Corsewall Point to Mull of Galloway ; its extreme breadth is about 5½ miles. The isthmus consists of a low, flat plain lying between Loch Ryan and Luce Bay. It has an average height of 70 to 100 feet above sea-level, sloping gradually to either beach. Piltanton Burn, which cuts this plain near its centre, forms a broad alluvial flat at Genoch House. Numerous hollows occur, most of which contain deep lochs.

As in the Machers, the highest land is on the west side. Indeed the whole peninsula may be looked on as exhibiting a long range of precipitous cliffs on the west, and sloping gradually to the sea on the east. Several peaks range about 500 feet, the highest being Craigenlee (592) in Leswalt, and Cairn Pyot (593), in Portpatrick, the greatest elevation in the Rhinns. With the exception of these rocky hills and the Moors of Galdenoch and Larbrax, most of the northern half of the peninsula is under cultivation. Towards Portpatrick the surface consists largely of drained land reclaimed from moor. In the southern half Barncorkrie Moor and Grennan Moor are still in a state of nature ; but with these exceptions most of the district has been brought under the plough. A prominent depression extends across the peninsula from Port Logan Bay to Terally Bay ; another forms a hollow between Clanyard Bay and Kilstay Bay ; a third connects the headland of the Mull of Galloway with the rest of the peninsula.

5. Rivers and Lakes

The main river system flows from N.W. to S.E. in long straight courses, with unimportant deflections. The chief streams are the Nith, the Urr, the Dee, the Fleet and the Cree. These are the oldest streams of the district, and an extraordinary fact is that some flow right across the elevations of the land. In some cases the valleys are longer than the streams. For example, the Dee rises near Loch Doon and flows S.E. past Castle Douglas, but the Dee valley is continued north towards Ayrshire, where it is occupied by the Doon, a river flowing to the N.W. Another remarkable point is that these streams run across the grain of the rocks. So it cannot be that the presence of soft belts of rock has determined their present channels. The valleys have a number of tributaries which converge towards them from opposite sides. The Black Water of Dee is one of those tributaries which are oblique to the course of the main stream ; the Palnure Burn is another ; and the Bladnoch also comes in as a tributary stream. A second class of streams has a course at right angles to the first group. They flow as a rule in accordance with the main slopes of the country, and follow the strike of the rocks. The Solway Firth, which is a drowned valley, has the same inclination, namely, from N.E. to S.W.

The main watershed is from N.E. to S.W., the highest ground running from Craigarie Fell to Mount Merrick, the Kells range, and on to the Windy Standard. But owing to the peculiar history of the river systems, the main valleys are cut right across this, and the actual watersheds more or less closely follow the same direction. The county margin from Darngarroch Hill runs for a considerable distance approximately on the watershed between the Nith (which receives a comparatively small part of the drainage of the county) and the Dee.   From the Windy Standard the county border crosses to the Doon Valley, and the eastern part of Kirkcudbrightshire belongs to the basin of the Urr.   The watershed between the Nith and the Urr starting on the shore near Southerness runs through Criffel, crosses the railway near Hills Tower, and swings to the west to the Nine Mile Bar, and thence to Darngarroch Hill, after which it follows the county line.   The watershed between the Urr and the Dee begins near Barcloy Hill, north of Dundrennan, and passes to the east of Castle Douglas, where the streams are only about four miles apart.   Then it runs N.W. towards Black Craig, reaching the county boundary about Trostan Hill.   The Dee valley is separated from the valley of the Doon at the county margin.   The watershed between the Dee and the Fleet is low and irregular.   In the S.E. it passes Fuffock Hill and Loch Whinyeon, goes through the White Top of Culreoch, past Loch Grannoch to Cairnsmore, by Gatehouse-of-Fleet Station, down Pibble Hill and Cairnharrow.   The watershed between the Black Water of Dee and the Cree is well defined.   Loch Dee drains to the Dee, Loch Enoch to the Doon, and Loch Trool to the Cree ; so that the watershed runs in an irregular manner among these

B

Head of Loch Trool

lochs. It ascends the Merrick and is continued north through Kirriereoch Hill.

The valley between the west side of the Cree and the Bladnoch is low and flat. On the west side of the Bladnoch valley there is a broad range of flattish ground occupied by numerous lochs and by large peat mosses. The watershed here winds out and in between the heads of the stream, passes through Carsecreugh Fell, and sweeps round to Quarter Fell. The N.W. side of the Luce valley is formed by a well characterised group of hills, of which Mid Moile is the most prominent. Towards the west this watershed passes through Glenwhan Moor. The Piltanton Burn, the only important stream in the Rhinns, rises to the west of Loch Ryan flows parallel to its shores as far as Lochan, then swings to the east to break through the sandhills flanking the Sands of Luce at their eastern extremity.

The Nith, which rises in Ayrshire some nine miles south of Cumnock, is joined by the Cluden Water at Lincluden, a mile and a half from Dumfries. " Lonely Cluden's hermit stream " is formed by the union of the Cairn Water and the Old Water of Cluden. The Cargen issues from Lochrutton Loch to join the Nith $2\frac{1}{8}$ miles south of Dumfries, and New Abbey Pow, after an eastward course of six miles, falls into it where the parishes of Troqueer and New Abbey march.

Issuing from Loch Urr, the river Urr is at first uninteresting and flows over an irregular channel. Its course from the Old Bridge of Urr is among level and well-cultivated grounds with a rich sward of grass. It enters Rough Firth at Palnackie, almost midway between Nith and Dee. Of its numerous feeders the only one of any importance is Kirkgunzeon Lane, which rises at Lang Fell and after a run of eight miles through lands largely alluvial, falls into it as Dalbeattie Burn, about a mile south of Dalbeattie.

The Water of Ken rises between Blacklorg Hill and Lorg Hill, and 17 miles nearer the sea enters Loch Ken —no loch at all, but merely an expansion of a sluggish river dreaming along between widespread lonely banks. At the southern extremity of the parish of Kells, 21 miles, from its source, the Ken is joined from the west by the Dee. From this point to the sea it passes under the name of its usurping tributary. The streams which feed the Ken are numerous but, severally, inconsiderable. On the left bank, midway between Dalry and New Galloway, it receives the romantic Garpel Burn, with its picturesque waterfall, the Holy Linn. Its principal tributary is the Deugh on the right bank, which, rising in three headwaters in Ayrshire, almost bisects Carsphairn, draining in two main divisions the whole of that extensive parish. Joined by the Polmaddy Burn, which has flowed eastwards from the slopes of the Carlin's Cairn, it pours the united waters into the Ken.

Of the ten or twelve rills which form the source of the Dee, the principal is the March Burn, which rises on the south-west slopes of Corscrine Hill (2668), changes its name to Sauch Burn, and then as Cooran Lane receives the surplus waters of Loch Dee. Thenceforward it is known as the Dee—the dark stream—or by its duplicate

name, the Black Water of Dee.   The dark colour of its
waters is due to the mosses among which it has its origin
and through which in its upper reaches it flows.   It is
worthy of note in passing that its salmon are said to be
of a darker colour than those of other rivers in the south
of Scotland.   Its course for 19 miles is in the main
south-eastwards.   It traverses Stroan Loch two miles
before its union with the Ken ; and from the confluence for five miles it expands into what is sometimes
called a second Loch Dee, a series of three successive
lakes with an average breadth of a quarter of a mile.
Its course now is rapid : a turbulent mill-race, it rushes
over a rocky bottom and between steep copse-clad
banks past Threave Castle Isle and Lodge Isle to Tongland, where at the Doachs it pours over a declivity of
rocks in an impetuous cataract.   Immediately below
Tongland Bridge, according to tradition, is the spot
described by the Scottish poet Alexander Montgomerie
(born about 1545) in the lines :

> " Bot, as I mussit myne alane,
>   I saw ane river rin
>   Out ouir ane craggie rock of stane,
>   Syne lichtit in ane lin,
>   With tumbling and rumbling
>   Amang the rockis round,
>   Dewalling and falling
>   Into that pit profound."

Three miles farther down it sweeps past Kirkcudbright,
and after five miles loses itself in the Solway.   Mussels
containing pearls of considerable value are occasionally
got in this river.   Anstool Burn from Balmaghie and

Glengap Burn, flowing out of Loch Whinyeon unite to form Tarff Water, the chief tributary of the Dee ; which, after a run of eight miles, it joins near Compstone House. About the middle of its course there is a picturesque succession of waterfalls, the Linn of Lairdmannoch, between 50 and 60 feet in height.

The Fleet, throughout a boundary river, is formed by the junction of two main streams, the Big and the Little Water of Fleet. The former has its head waters in three burns which rise on the eastern slopes of Cairnsmore of Fleet. One of these, the Carrouch Burn, divides Anwoth from Kirkmabreck ; the Big Water, and thereafter the united streams, divide Anwoth parish from Girthon. Issuing from Loch Fleet, the Little Water flows south to join the Big Water just above Castramont. Wild, heath-clad hills overlook the upper part of its course, while its middle and lower reaches are flanked by declivities and plains, here richly wooded and there stretching backwards in well-tilled fields. A mile below Gatehouse the river suddenly expands into an estuary $3\frac{1}{2}$ miles long and a mile in average breadth.

The Cree is a boundary river. It has its source in Loch Moan, and for several miles flows through a bleak moorland district separating the Stewartry from Ayrshire. Opposite the north end of Loch Ochiltree it bends sharply to the east for over a mile, and then, for the remainder of its course, flows south-eastwards between the Stewartry and the Shire. Near the farm of Brigton it is joined by its chief tributary, the Minnoch, reinforced by the Water of Trool. For three or four miles below this it flows with an almost imperceptible current through a broad channel known as the Loch of Dee. On the left bank, through the beautiful Linn of Cadorcan, the waters of Cadorcan Burn fling themselves in a lovely cascade into the Cree over a cliff

The Cree at Machermore

some fifty feet high in the Wood of Cree, one of the few remaining fragments of the ancient forests of Galloway. Right across were the Cruives of Cree, where salmon used to be caught in traps formed of stakes and wattles fixed to a chain stretched across the river. The Cruives of Cree find a place in what is probably the oldest form of the lines proverbial of the power of the Kennedy family in the sixteenth century.

> " 'Twixt Wigtoune and the Toune of Aire,
> And laigh down by the Cruives of Cree,
> Ye shall not get a lodging there,
> Except ye court a Kennedie."

Augmented by the Penkill Burn, which joins it just above Newton Stewart and by the Palnure Burn, which falls into it three miles above Creetown, the " crystal Cree " makes its way by a broadening estuary into Wigtown Bay. This is one of the very few Scottish rivers visited by that delicate fish, the sparling.

The Bladnoch flows out of Loch Maberry and, though with many windings, maintains on the whole a south-east direction to its mouth. Its main feeder, Tarff Water, rises on the slopes of Benbrake Hill, and flows almost parallel to it between New Luce and Old Luce on the west and Kirkcowan on the east, till six miles from the confluence of the two streams it swings to the north-east across the last-named parish.

Till within seven miles of the sea the Luce consists of two streams, the Main Water and the Cross Water of Luce. Both rise in Ayrshire ; in their higher reaches both flow through bleak moorlands, and both are augmented by numerous brawling burns. At the village of New Luce the Cross Water strikes the Main Water at right angles, and from this point the Water of Luce makes for Luce Bay, which it enters through a small estuary, dry at low water.

Galloway yields to no district in Scotland in the number and beauty of its inland waters. The glaciers which streamed southwards scooped out hollows in the

Silurian rocks, many of which remain to-day as lochs. As a rule they are small, and nearly all contain islands. Of over forty in Wigtownshire and thirty in Kirkcud-brightshire, the largest is only two miles long. They occur singly and in groups ; they are met with at almost the level of the sea and at elevations ranging to 1700 feet. They are in general well supplied with fish ; Lochs Grannoch, Doon and Dungeon contain char ; while tailless trout are the boast of silver-sanded Loch Enoch.

Loch Rutton, 325 feet above sea-level, supplies Dumfries with water. Near Craigend Hill are the romantic loch of Lochaber and Loch Arthur, so named from the tradition of King Arthur's sojourn in the vicinity. A mile from the Solway, Loch Kinder, blue in the hollow of Criffel, no longer supplies chairmakers with bulrushes and weavers with reeds. Loch Urr is a picturesque sheet of 106 acres lying in the moorland. Between the parishes of Kirkpatrick-Durham and Urr is Loch Auchenreoch, and a mile to the east, Loch Milton. Loch Dee, 253 acres, is an irregularly shaped lonely mountain lake in a treeless waste near the Dungeon of Buchan. About a fifth larger, embosomed among rugged hills and solitary moorlands, is Loch Grannoch, the best trouting loch in Galloway. Loch Skerrow, 125 acres, has five or six islets wooded with birch and alder. The lochlet of Lochanbrek, at an altitude of 650 feet, is near a spa formerly much resorted to. Loch Dungeon, at a height of 1025 feet, is flanked by steep hills on the south and rugged crags on the west.

Loch Ken, 4½ miles long, and from 200 to 800 yards

wide, is the largest loch in the Dee basin. Flanked on the west side by a range of hills, which on the north and centre press close upon its edge, and at its southern corner terminate in a huge rock, its shores are here and there fringed and tufted with plantations. Its surface is broken by four beautifully wooded islets. Carlingwark Loch, 105 acres, formerly much larger, was partially

Carlingwark Loch, Castle Douglas

drained in 1765 for the purpose of procuring marl for manure. Near it stood the Three Thorns of Carlingwark, for ages a trysting-place of laird and yeoman in Galloway. Loch Whinyeon, 700 feet above sea-level, had its waters diverted about a hundred years ago from the basin of the Dee, to which it belongs, to drive the cotton mills of Gatehouse-of-Fleet. From the south-east corner of Loch Fleet, about a mile east of Loch Grannoch, issues the Little Water of Fleet. Loch Enoch (said to be a corruption of Loch in Loch, from one of its islands having a loch in it), at an elevation of 1650 feet is a veritable lake in cloud-land. Loch Neldricken has at its edge an emerald stretch of reeds, in the middle of which is a circular expanse of deep black water. It never freezes, say the natives, not even in the bitterest winters ; and it bears the significant name of the Murder Hole.

Loch Valley is a fine example of a moraine-formed

The Murder Hole, Loch Neldricken

lake ; it is surrounded by numberless boulders and perched blocks, and rocking stones, many of them so exquisitely poised that a light breeze disturbs their equilibrium. Among the highest mountains of Galloway, its shores steep, rugged, and wooded, lies Loch Trool with an undulating beach which, by two constrictions, divides it into three distinct basins. Its extensive drainage area includes the southern slope of the Merrick and the northern of Lamachan. At the end of the loch is the finest waterfall in Galloway. Buchan Linns

One of the Buchan Falls, Glen Trool

have been formed by Buchan Burn cutting a deep gorge between two hills. Through this it hurls itself by a succession of leaps into the lake 120 feet beneath. The district is rich historically. It witnessed stirring scenes in the Brucian struggle for Scottish independence, and its hills and corries were familiar with the struggles of the Covenanters.

On the county march are Lochs Maberry, Ochiltree and Dornal—the last belonging more properly to Ayrshire. On one of the eight islets of Maberry are the remains of an old castle. The lochs of Mochrum are seven in number ; Castle Loch drains into Mochrum Loch, the largest in the basin. Lochs Magillie and Soulseat are within easy access of Stranraer. The latter, surrounded by trees, is almost bisected by a peninsula which projects into it. Here stood the now-vanished Monastery of Soulseat. Other lochs in the basin of the Luce, all near the eastern shore of the bay, are Whitefield, Eldrig, and the White Loch of Myrton. In Loch Ryan basin are the White and Black Lochs of Inch, connected by a canal. The space between them—the " dressed grounds " of Castle Kennedy—is laid out in formal terraces and alleys. Avenues of coniferous trees, beds of flowering plants and shrubs, ponds bedecked with waterlilies, are features of a piece of landscape-gardening unexcelled in the south of Scotland.

6. Geology

Geology is the science that deals with the solid crust of the earth ; in other words, with the rocks. By rocks, however, the geologist means loose sand and soft clay as well as the hardest granite. Rocks are divided into two great classes—igneous and sedimentary. Igneous rocks have resulted from the cooling and solidifying of molten matter, whether rushing forth as lava from a volcano, or, like granite, forced into and between other rocks that lie below the surface. Sometimes pre-existing rocks waste away under the influence of natural agents as frost and rain. When the waste is carried by running water and deposited in a lake or a sea in the form of sediment, one kind of sedimentary rock may be formed—often termed aqueous. Other sedimentary rocks are accumulations of blown sand : others are of chemical origin, like stalactites : others, as coal and coral, originate in the decay of vegetable and animal life. Heat, again, or pressure, or both combined, may so transform rocks that their original character is completely lost. Such rocks, of which marble is an example, are called metamorphic.

Examining the order in which rocks occur, the materials which compose them, and the fossils or petrified remains of plants and animals which they contain, geologists have arranged groups of rocks according to their relative age. Lowest of all are the Archaean rocks. Then in order come (1) rocks of ancient life, or Palaeozoic ; (2) rocks of middle life or Mesozoic ; and (3) rocks of recent life, or Cainozoic. The following table shows the usual classification of Palaeozoic stratified rocks, the youngest on top.

> *Permian*   *System.*
> Carboniferous   ,,
> Old Red Sandstone   ,,
> Silurian   ,,
> Ordovician   ,,
> Cambrian   ,,

The oldest rocks exposed in Galloway are Lower Silurian or Ordovician. These form a broad strip of country from Sanquhar past the Merrick into Loch Ryan and the Rhinns. The Upper Silurian rests upon these conformably; its outcrop lies to the south-east, forming the whole of the country from the Mull of Galloway to Dumfries, with the exception of a narrow coastal belt. The outcrop of the Upper Silurian is about 21 miles broad from Dalry to Kirkcudbright, and the outcrop of the Lower is about 16 miles broad from Dalry to the foot of Loch Doon. One striking feature of these rocks is that the beds of strats are very steep. This is due to disturbances which the rocks have undergone. Careful observation proves that the same beds are repeated many times in any good natural section such as a stream-side or road-side. This is a natural consequence of folding. The beds were deposited as flat sheets of mud and sediment. The folding is like what takes place when the bellows of a camera are shut up. The individual folds are sometimes vertical, though very often inclined. When the folding is inclined and the two sides of the arches and troughs are parallel it is said to be isoclinal. This is the great characteristic of the whole of the district. The folds have a common extension or *strike*, which in the whole Southern Uplands of Scotland points from S.W. to N.E. The main streams, which run from N.W. to S.E., cut across the folding structure of the country. The present system of the ground depends upon erosion. In all arches or *anticlines* the lowest rocks form the interior or core ; while conversely, in the troughs the lowest rocks form the exterior. When an arch has had its top cut away by denudation the underlying rocks are exposed in the centre of the arch. In many places the Lower Silurian rocks, one of which is a very characteristic hard radiolarian chert, have been exposed in this manner in the midst of Upper Silurian rocks. On the map these show as boat-shaped outcrops.

The Silurian rocks are probably several thousand feet thick. A traveller crossing the Uplands would obtain the impression that they were much thicker than they really are. This is misleading : the rocks are being repeated every few hundred yards. The Lower Silurian comprises a series of volcanic rocks or lavas belonging to the Arenig sub-division. These, which cover no large area within the counties, appear only here and there in the cores of folds. Over them lie black mudstones and radiolarian cherts. The latter, which are flinty grey-green or red rocks, very hard and splintery, when examined under the microscope are seen to consist of the shells of radiolaria. Outcrops of the Chert series are comparatively frequent, but nowhere large.

The next sub-division of the Lower Silurian is called

the Llandeilo-Caradoc. It includes greywackes and shales, some of which are black and contain many graptolites. Two of the best known of these bands are the Glenkiln Shale and the Hartfell Shale. The Upper Silurian lowest division is known as the Llandovery Taranion. It consists also of greywackes, mudstones and shales, and it contains one well-known graptolite-bearing band, the Birkhill black shale. The highest rocks, the Wenlock and Ludlow, form a narrow belt to the south of Kirkcudbright (also the south end of Burrow Head), and on to the mouth of the Nith. While these rocks were being deposited this district was occupied by a compact shallow sea, in which graptolites flourished together with molluscs and brachiopods ; but no fish remains and no plant remains are preserved in these strata, and it is doubtful whether as yet fishes were in existence. The sequence of the Silurian rocks in Kirkcudbrightshire and Wigtownshire is not yet complete, the topmost members, the Downtonian, being missing. Then followed the folding and crumpling of the Silurian strata which, up to that time, had been flat. This folding was the result of great earth movements which took place over a very large part of the west of Europe. The compression took place in a N.W. to S.E. direction, and hence the uniform strike of the folds. After the folding was completed the sea bottom was upheaved and formed into dry land, and the process of erosion began, which has continued unbroken ever since.

The epoch of folding was followed by intrusions of

granite. These are generally assigned to the Lower Old Red Sandstone period. During this time there were great chains of volcanic mountains over the south of Scotland, of which the Cheviots and the Carrick Hills are well-preserved fragments. The granite masses of Galloway include the Merrick mass, the Cairnsmore of Fleet mass, and the Dalbeattie mass, each of which is 10 to 12 square miles in area. Smaller masses occur at Creetown, at Crummag Head and elsewhere. Several small patches of dark-coloured granite containing hornblende (diorite) are to be met with, as at Ardwell, at Eldrig village and at Culvennan, some 3 miles north of Kirkcowan. The granite masses rose into position in a state of fusion, intensely hot, and the rocks in contact with the granite were profoundly altered and re-crystallized. For example, at New Galloway the Silurian shales and grits have been changed into mica schists, which contain sillimanite and other contact minerals produced by high temperatures. The granite when in a liquid state had burst through the rocks, sending veins and dikes into their fissures. The granite never reached the surface, but consolidated under a great overlying mass of rock, which has now been swept away. It is possible, however, that the granite formed the centre of volcanoes of which no trace now remains. Even at great distances from the granite numerous dikes are found cutting through the Silurians. A swarm of these occurs between Castle Douglas and Kirkcudbright. Though not broad, many of them run for a long distance.

C

It is likely that during the Old Red Sandstone period all Galloway was dry land ; there are no Old Red Sandstone deposits now preserved anywhere in it.   In the next succeeding period, the Carboniferous, Galloway was at first a range of hills, while the centre of Scotland and the north of England were covered by the sea. In course of time Galloway became an island, which gradually sank lower and lower.   The sea finally rose and flooded its valleys, in some of which deposits of carboniferous rock were formed.   A strip of such rock occurs at Abbeyhead and again at Kirkbean.   These belong to the lowest part of the Carboniferous deposits and are known as calciferous sandstones.   On the west side of Loch Ryan also there is a belt of Carboniferous rocks ; these are of considerably later age and belong to the Coal Measures period.   After the Carboniferous period ended, dry land again supervened, and the red sandstones of Maxwelltown and the breccias of Loch Ryan were subsequently deposited, possibly in desert lakes.   These sandstones contain the footprints of reptiles, but no other trace of life.   At Loch Ryan they rest on Carboniferous, but at Maxwelltown they rest directly on the Upper Silurian.

For a very long period the geological history of Galloway is a blank.   In early tertiary times many long dikes of basalt were injected into the Silurian rocks.   A few instances occur in Galloway, *e.g.* at Kirkcolm. During the glacial period Galloway was buried under a deep layer of ice.   Great masses of snow accumulated on the high hills and formed a moving ice-sheet, which streamed southwards into the Solway, carrying with it numerous blocks of rock, which were deposited along its course. Blocks of Criffel granite are found near Birmingham and in South Wales. The Firth of Clyde was filled by a great ice-stream coming down from the Highlands, and this passed into the Rhinns. After the

Loch Valley

(A moraine-formed lake)

main ice-sheet melted, local glaciers existed in the high hills, where glacial moraines are still conspicuous features of the landscape. Abundant evidence of the Ice Age in Galloway is to be met with in scratched rock surfaces, boulder clays, sands and gravels and the erratic blocks just mentioned. A very striking feature all over Galloway is the manner in which the boulder clay has been deposited. It is found in large, smooth ridges, oblong or rounded in shape, locally known as " drums."

A small patch of blown sand occurs beside Port Logan ; another at the Point of Lag, where it rises into a hill 75 feet above the sea ; and a larger strip at the head of Luce Bay, stretching from Sandhead to the mouth of Piltanton Burn.

In the Stewartry parts of Irongray, Terregles and Troqueer have the soil a sandy loam. A belt stretching from Maxwelltown along the shores of New Abbey and Kirkbean has a soil either of carse or rich loam with a subsoil of gravel or limestone. In the south-east of the country and in the valleys of the Ken, the Fleet and the Cree, a dry loam of a hazel colour is met with. In the upland districts the soil, as a rule, is thin and mossy.

In Wigtownshire along the side of the lower reaches of the Cree and at the head of Wigtown Bay the soil is alluvial. In much of the Machers and a large portion of the Rhinns it is a dry hazelly loam, as also in the cultivated part of the moors. In the centre and north of this division great tracts are covered with peat moss resting sometimes on a bed of marl, though frequently on a substratum of clay.

7. Natural History

In recent times—recent, that is, geologically—no sea separated Britain from the Continent. The present bed of the North Sea was a low plain intersected by streams. At that period the plants and the animals of our country were identical with those of Western Europe. But the Ice Age came and crushed out life in this region.

In time, as the ice melted, the flora and fauna gradually returned, for the land-bridge still existed. Had it continued to exist, our plants and animals would have been the same as in Northern France and the Netherlands. But the sea drowned the land and cut off Britain from the Continent before all the species found a home here. Consequently, on the east of the North Sea all our mammals and reptiles, for example, are found along with many which are not indigenous to Britain. In Scotland, however, we are proud to possess in the red grouse a bird not belonging to the fauna of the Continent.

While displaying the general flora of Scotland, Galloway, from its position, shares in the plants characteristic of the west and the south.

Of flowering plants there are over 900 species in Galloway ; of ferns over 20 species. The moss flora is exceedingly well represented, especially in Kirkcudbrightshire ; liverworts, lichens and fungi flourish wherever the conditions are favourable.

In the farthest-out rock pools at the lowest of low water one finds the edible *Alaria esculenta* or honeyware, and the familiar *Laminaria digitata* or tangle. Somewhat nearer the shore in some localities may be met with *Odonthalia dentata* and *Chondrus crispus*, or Irish moss. These are red algae. Still nearer the shore are several species of *Fucus* or bladder wrack and of *Polysiphonia*, and here and there *Himanthalia lorea* or sea thongs, all of which are olive algae. In the pools nearest the shore one gathers the beautiful grass-green *Ulva latissima* or laver, *Enteromorpha compressa* and various species of *Cladophora*, all green algae.

The flora of the coast is determined largely by the nature of the soil. On rocks one finds, but rarely, *Crithmum maritimum* or sea samphire, and sea campion and michaelmas daisy in abundance. On sandy shores, but very rare, are purple sea-rocket and sea holly ; and halberd-leaved orache, prickly sea-weed, sea kale, thyme-leaved sandwort and sea purslane. Further inland rest-harrow, bird's-foot trefoil, yellow bedstraw and others occur ; while still further from the sea there are marsh arrowgrass, seaside arrowgrass, seaside plantain, sea milkwort, and scurvygrass. On muddy shores, and entirely submerged at high water one meets with broad-leaved grasswrack and glasswort.

Plants usual to river valleys are very numerous, and the lake-side flora is also rich and varied. In mosses cross-leaved heath, common ling, bog myrtle, bog asphodel, cranberry and sundew are abundant. In sub-alpine districts are to be found large-flowered bitter cress, giant bell flower, and many others. Higher up the mountain sides are alpine meadow rue, least willow, wild thyme, cotton grass and juniper. Parsley fern is plentiful on the higher hills, Wilson's filmy fern is common in sub-alpine glens, and moonwort, adder's tongue, hart's tongue and, very rarely, the royal fern are also to be got.

In 1905 Kirkcudbrightshire had 19,708 acres in woods and plantations, or roughly $\frac{1}{30}$th of the area, but at one time the greater part of the county was covered with wood, largely oak. This is shown by place names, remains of natural timber on the sides of hills and banks of rivers, and by the numerous peat mosses out of which trunks of trees are still dug in good preservation. Of ancient forests may be named the forest of Minnigaff, the Free Forest of Cree, the Forest of Buchan in Kells, the Forest of Kenmure, the (small) Forest of Rerwick, the Forest of Colvend and the Bishop's Forest in Iron-gray. In those days wood was the common fuel ; and, in addition, much was consumed by the saltpans along the coast. Wigtownshire had, in 1905, 8526 acres of woodland.

The littoral fauna of Galloway comprises those animals which are to be met with from high water mark to a depth of, say, 25 fathoms, though denizens of the further deep now and then visit the coast. Many causes combine to make the shore life of Galloway varied and abundant. The Nith, the Urr, the Dee, the Cree and the Luce have estuaries with extensive mud flats. Here are long stretches of sand, and there the bottom is rocky. Currents coming south through the North Channel and sweeping north through St George's Channel and the Irish Sea, bring with them animals from the Boreal and the Lusitanian regions. The abundance of fresh-water organic matter brought down by the rivers and smaller streams helps to swell the supply of food.

The crumb-of-bread sponge is common on the stems of oarweed, and in crevices of rocks. Zoophytes and sea firs are plentiful. Several species of sea anemone flourish between tide marks where the shore is rocky.

On sandy stretches one comes upon terebella, or the sand mason, its long tentacles and containing-tube plastered with sand and shell and stone. Where the sand holds much organic matter, the lob-worm is usually present in numbers. The common starfish and the common brittle star are abundant ; the sea urchin is frequent far out among the oarweeds ; and in some places after a storm the shore is white with the tests of the heart urchin. The shrimp, the lobster, the edible crab and the shore crab are found, as is also the hermit crab with its companion the beautiful Nereis worm within its protecting whelk or buckie shell. Whelks and mussels form articles of commerce. The Bay of Luce is noted for its razor-shells, and the oysters of Loch Ryan have more than local fame. Of fishes, the saithe, the lythe and the skate are plentiful at certain seasons. At times the coast is visited by shoals of mackerel. The father lasher and the grey gurnard are common, and in spring the lumpsucker comes to the shore to spawn.

Cod is plentiful, haddock somewhat less so ; while halibut, though occasionally got, is not common. The plaice, the dab and the sole are very numerous, and the sparling is a valuable fishery in certain tidal rivers in winter and spring. Anchovies are not unknown ; in 1889 the Bay of Fleet was alive with shoals of them. The principal fresh-water fish are the salmon, the trout, the perch and the pike. Of aquatic mammals the porpoise and the grampus are frequent visitors ; a dolphin is now and then captured by stranding or otherwise, and sometimes a school of whales is driven ashore.

Of reptiles one may name the adder, the lizard and the slow-worm; of amphibians the frog, the toad, the smooth newt, the crested newt, and up on the hills among moss hags the palmated newt.

Where there are suitable woods the roe-deer is frequent in the Stewartry, less so in Wigtownshire; and fallow-deer are to be seen in parks in a more or less domesticated state. On the upper hills the alpine hare is well established, and everywhere the rabbit, the brown rat and the house mouse are more numerous than is to be desired. The watervole is frequent; and the ravages of the short-tailed field vole some twenty-one years ago are matter of history. The fox issues from thick copses or descends from the higher hills to pursue the depredations which render him offensive to shepherd and game-keeper. In stream and lake the otter carries on his fishing. Weasel and ermine, mole and hedgehog are very general.

Galloway is not rich in bats, but bird life is very abundant. The rook, the raven and the carrion crow occur in considerable numbers. The magpie is not common. The starling, the green-finch and chaffinch are in great profusion. The goldfinch is not nearly so numerous as it used to be, though of recent years there is a tendency to increase; the bullfinch is common enough in woods and gardens. Several species of buntings and of wagtails are found. The skylark showers down floods of silver melody as it soars over fell and moor and green mountain. Robins, pipits and tits are common. The blackbird whistles in many a garden croft, and on many a bush the wise thrush sings each song twice over. Owls are plentiful. The barnacle goose occurs in immense numbers, and the wild duck is very abundant. The grouse moors of Wigtownshire are among the best in Scotland ; while black cock and snipe, partridge and pheasant afford sport to many a gun. Of the numerous shore birds we must note the oyster-catcher, the golden plover, the dunlin and the ubiquitous gull.

8. Along the Coast

From Cargen Pow at the head of the long and gradually broadening estuary of the Nith to Creetown at the head of Wigtown Bay, the coast of Kirkcudbrightshire is about 60 miles in length. It is broken into by four expansions of considerable size, the Rough Firth, Auchencairn Bay, Kirkcudbright Bay and Fleet Bay. At Aird Point, 4 miles south of Cargen, the Nith enters the Solway Firth, and here the sea-board begins. Rather more than a mile inland are the picturesque ruins of Sweetheart Abbey. Clayey and low, the New Abbey shore is flanked by merseland which forms excellent pasture. Bending slightly to the west the shore passes the mouth of Abbey Pow, and for a little over 3 miles runs almost due south as far as the village of Carsethorn. Rounding Borron Point and passing the ruins of M'Culloch's Castle, we reach Arbigland, where in 1747

John Paul, better known as Paul Jones, the famous sailor, was born. The coast at this point is precipitous, and there are some very singular rocks, notably a natural Gothic arch called the " Thirl Stane." But with the exception of these and a few low rocks at Satterness,

Medallion of Paul Jones

the shore as far as Southwick Burn is low and sandy, with here and there belts of links gained slowly from the sea. At Satterness is the oldest lighthouse in Galloway. At one time there were salt pits here, and from these comes the name Satterness, the etymology of which is lost in the present-day Southerness.

A sharp turn westward and 4 miles bring us to the mouth of Southwick Burn, beyond which begin the

" wild shores of caverned Colvend." Chief of the

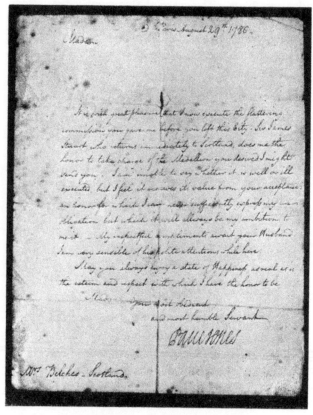

Facsimile of Letter of Paul Jones

caverns is the Piper's Cove, 120 yards in length, with a well in the middle 22 feet deep. Here too is the

singular arch in the cliff known as the Needle's Eye.
Between Douglas Hall and Urr Waterfoot, at the

The Needle's Eye, Douglas Hall

entrance to Rough Firth, a range of reddish-lichened
copse-clothed cliffs rises to a height of 200 feet at Castle
Hill of Barcloy and 400 feet at White Hill.

At Rockcliffe the shore is rocky with wide stretches of hard, smooth sand.   Kippford, locally known as the Scaur, is a fine watering-place.   The seaboard of Buittle consists of a peninsula running $2\frac{1}{2}$ miles down to Almorness Point, washed on the east side by Rough Firth and on the west by the bays of Orchardton and Auchencairn. Near the former is Orchardton Round Tower, the only one of its kind in Galloway.   It was generally supposed to have been built as a stronghold by Uchtred, Lord of Galloway, in the twelfth century, but Train recognised in it " a fine specimen of the Danish rath," while modern experts attribute it to the fifteenth century.   Lying about midway between Almorness Point and the Point of Balcary is Hestan Island, the Isle Rathan of Crockett's *Raiders*.   From Balcary to the mouth of Dunrod Burn the trend of the coast is roughly W.S.W.   For the most part bold and ironbound, it presents a series of abrupt headlands, 100 to 350 feet high, and is intersected by the baylets of Rascarrel, Barlocco, Orroland, Port Mary, Burnfoot and Mullock.   At various points occur caves which have been drilled in the cliffs by the ceaseless action of the sea.   At Barlocco the Black Cove, 265 feet long, 90 wide and 40 in height, and the White Cove, 252 feet by 90 (at its widest) by 60, are particularly noteworthy.   In recent years they have gained an added interest from the use made of them in Crockett's *Raiders*.   At Port Mary is shown a granite boulder from which Queen Mary of Scots is said to have stepped into the boat which was to carry her to the Cumberland coast.

From Mullock Bay to Torrs Point the coast is on the whole rocky. In a precipice on the Balmae shore is a remarkable cavern, Torrs Cove, running some 60 feet

Cave, Rascarrel

into the rock. Narrow at the entrance and then gradually widening, it rises near the middle to a height of fully 12 feet, after which it contracts towards the farthest end. Kirkcudbright Bay, which may be said to begin with the precipitous cliffs of Torrs Point, runs northwards for about 4 miles, with a breadth varying from 1 mile at the entrance to 2 miles at its widest. The well-wooded peninsula of St Mary's Isle divides the upper part of the Bay into two, forming on its eastern side the Manxman's Lake, the principal anchorage in the Bay. On the western side is the estuary of the Dee,

> " King of all the streams
> That roll to Scotland's southern sea."

On St Mary's Isle stood a priory founded in the reign of David I by Fergus, Lord of Galloway. A beautiful walk down the west side of the Bay leads past the " Auld Kirkyard of Kirkchrist," the Nunmill, where an old archway indicates the side of an ancient nunnery, and the old churchyard of Senwick, the burial place of MacTaggart, author of the *Gallovidian Encyclopaedia*. Past Balmangan Bay the Peninsula of Meikle Ross is reached, opposite which, and separated by a narrow strait, is the Little Ross Island. Rounding the Ross, we come to the wide expanse of Wigtown Bay. Fallbogue Bay and Brighouse Bay are passed in turn ; then Borness Point, with its wave-worn cliffs crowned by the remains of an ancient fort, known as Borness Batteries, and its Bone Cave, the exploration of which has proved of great archaeological interest. For the rest of its length the trend of the coast is to the northwest, the only break of any size being Fleet Bay. In the little churchyard of Kirkandrews is buried William

Nicholson, the greatest Galloway poet, and author of the *Brownie of Blednoch.*

The Isles of Fleet, Barlocco Isle, Ardwall Isle and Murray's Isle, lead up to the mouth of Fleet Bay. The bay is flat and sandy, and the shores low. Rather more than a mile from the mouth of the Fleet is Gatehouse, picturesquely situated on both banks of the river. From Gatehouse to Creetown has been described as " perhaps the most beautiful shore-road in Britain." And indeed for beauty of scenery, hill and valley, moorland and shore, " Fair Anwoth by the Solway " is unrivalled in the south of Scotland. Just after passing Cardoness Castle, on the west shore of the bay, we catch sight of Rutherford's Monument. Then comes Ardwall House, and next Skyreburn Bridge. The water in Skyreburn and similar streams often rises with surprising and unexpected suddenness. Hence the proverb, " A Skyreburn warning," that is, no warning at all.

The coast near Ravenshall and Kirkdale is rugged with steep cliffs rising to a considerable height, in some cases perpendicular to the sea. But with this exception the Kirkmabreck shore is flat, sandy and shelly. Here and there the cliffs are pierced with caverns, the most notable of which is known as Dirk Hatteraick's Cave. About a mile from Ravenshall is Kirkdale House, near a romantic glen of the same name, while a short distance along on the opposite side of the road are the ruins of Carsluith Castle. A mile before entering Creetown, the western extremity of the county coast, is the Mansion

D

House of Cassencary " finely situated in a level holm studded with trees."

At high water vessels of sixty tons ascend the Cree as

Rutherford's Church, Anwoth

far as Carty, some 3½ miles above Creetown ; but it is at Balsalloch, opposite the " Ferry Toon " that the coast of Wigtownshire may be said to begin. Here the receding tide leaves bare a stretch of sand a mile broad, which increases to a breadth of a mile and three quarters at the mouth of the Water of Bladnoch, and then gradually narrows to its southern limit in Orchardton Bay, 5 miles further down the coast. Just before we reach the Bladnoch, Wigtown is passed, " the quaintest, auld farrantest county village in Scotland." A little south of the Bladnoch are the remains of the

The Gateway, Baldoon Castle

old mansion house of Baldoon, the scene of the death of the Bride of Lammermoor, " The dear, mad bride who stabbed her bridegroom on her bridal night."

From Orchardton Bay the coast trends eastward past Innerwell Point, and then south past the ruins of Eggerness Castle, where the shore becomes rocky, rugged, and picturesque. Eggerness (Edgar's Ness) Point overlooks Garlieston Bay with its trim little

village.   Near the village is Galloway House, long the
principal seat of the Earls of Galloway.   The trend is
now almost due south, and creek and cove, foreland
and cape, carry a bold and precipitous coast, pierced
here and there by deep caves, to the Isle of Whithorn,
and then south-west to Burrow Head.   About 2½ miles

Remains of Cruggleton Castle

south of Garlieston Bay is the site of what was once the
famous Castle of Cruggleton.   All that now remains is
an arch 10 feet high by 13 feet wide ; but from early
days to the close of the sixteenth century it was one of
the chief castles of note in Galloway.   At the Isle of
Whithorn in 396 St Ninian began his mission.   Two
miles to the south is the bold promontory of Burrow
Head, on the top of which are traces of a small fort
or cairn, an outlook station of the old sea-rovers·

Rounding this, we come in sight of Luce Bay. This huge sheet of water, covering an area of about 160 square miles, is 18 miles wide at the mouth and narrows to 7 miles along its northern shore, where the Sands of Luce run out for half a mile at low water.

After Burrow Head we pass the ruins of Castle Feather and of Port Castle, and reach Port Counan Bay, on the south side of which is St Ninian's cave. Here, tradition has it, the saint was wont to retire for meditation and prayer. The cave is 27 feet long and about 10 high. For the greater part of its length the Glasserton shore is backed by a chain of green-topped hills. Then a mile of steep cliffs is succeeded by an old raised sea-margin of smooth gravel with high grassy cliffs beyond. Monreith Bay with its beautiful scenery is followed by Barnsalloch Point, crowned by the remains of a fort, Danish or Anglo-Saxon according to the antiquary one consults. A mile and a half north of this is Port William. Sweeping round Auchenmalg Bay at a distance of 9 miles from Port William, we come upon the headland of Sinniness (Sweyn's Ness), not far from which are the ruins of Sinniness Castle. Farther on is the mouth of the Water of Luce, and Glenluce village with its stately Abbey ruins. The river mouth is flanked by level lands, while a broad fringe of sands, dry at low water, stretches right across the head of the bay. Here is the fishing village of Sandhead. Broken by a number of small bays, Chapel Rossan, New England and Drummore, the shore reaches East Tarbet about 9 miles farther south. Drummore village stands on

Drummore Bay. At Tarbet (Tarbert) two bays run inland from opposite sides till they nearly meet. Tarbet means " drawboat," and in bygone days it was the

St Medan's Chapel
(Near the Mull of Galloway)

custom to draw vessels across this narrow isthmus in order to avoid the dangerous tides of the Mull. From Tarbet the headland of the Mull stretches eastwards for a mile ; its extremity 210 feet high is crowned by a lighthouse. Its southern shore rises in cliffs over

200 feet high. From one of these, so the legend goes, the brave old Galloway chief was flung—*Ultimus Pictorum*—carrying with him the secret of heather ale.

With the exception of the Bays of Clanyard, Killantringan, Port Logan, Ardwell and Dally with their sandy beaches, the western coast of the Rhinns—the Back Shore—is bold and rocky; the cliffs here rising precipitously, and there ascending by grassy slopes. Fissures in the cliffs are numerous, and in many places there are caves with narrow openings but roomy interiors. Clanyard Bay is flanked by the ruins of Clanyard Castle; Port Logan Bay has on its north side a circular tidal fish-pond, one of the wonders of Galloway. Tradition says that a ship of the Spanish Armada was wrecked at Port Float. Port Spital suggests the former existence of a hospital or hospice. By the ruins of Dunskey Castle, we reach Portpatrick, the most popular holiday resort in Galloway. North of Killantringan Bay is the Kemp's Wark, name reminiscent of the days of the Northmen. At Saltpans Bay salt is no longer extracted from sea water, though the name persists.

From Dally Bay the land inclines to the north-east as far as Corsewall Point, which carries the ruins of the old Castle of Corsewall or Crosswell. Two and a half miles east of this Milleur Point is reached, and Loch Ryan is entered. The loch runs inland for eight miles, with a breadth varying from a mile and a quarter within the entrance to two miles and a half. For about three

miles from Milleur Point the coast resembles that of
the Back Shore, but opposite Kirkcolm village its
character changes.   A shelving bank of sand, the Scar,
projects south-east into the Loch for a mile and a half.

Dunskey Castle

Beyond this is The Wig, a fine natural basin, and thence
to Stranraer at the head of the Loch the shore is low
and sandy.   Stranraer is the chief centre of population
and commercial activity in the county.   The eastern
shore of the Loch is flat to Cairnryan village, and thereafter rocky and cave-pierced to the Galloway Burn,
where the Wigtownshire coast ends.

9. Raised Beaches. Coastal Gains and Losses. Lighthouses

At various elevations—from 10 to 150 feet—above the present lea-sevel there occur tracts of ground which have been sea-beaches in former ages. These terraces, known as raised beaches, have originated through successive slow risings of the land with long pauses between. The 25-foot beach can be seen with fair continuity along the western shore of the Bay of Luce, but never extending very far inland. On the opposite side of the Bay there is evidence of a terrace cut out of the boulder clay at a time when the land was 40 to 50 feet lower than it is now. Two fragments may be mentioned : one extending some three miles from Port Counan to Cairndoon, the greater part shingly, but cultivated at its north-west end ; and the other running northwards from Monreith. The low-lying undulating ground between Luce Bay and Loch Ryan for the most part is covered with sand and gravel deposited in terraces, the most noticeable of which forms the 25-foot beach. Along the shore of Luce Bay from Auchenmalg to Port William the 25-foot beach is distinctly traceable as a shelf of gravel extending inland from the present beach for 50 to 100 yards. It is seen at Garlieston Bay, at Orchardton and Baldoon, is well marked from Macher-more to Wigtown sands, and is easily traced from Creetown to Ravenshall, where it forms a belt of level ground between high-water mark and an older sea-cliff.

Indeed this 25-foot beach forms a prominent feature all along the southern shores. On the Fleet below Gatehouse, on the Dee between Tongland and St Mary's Isle, past Auchencairn, below Kirkennan on the banks of the Urr, from Caulkerbush to Southerness as a tract of carse land, past Carsethorn on to Ingleston and up to Kirkconnel, the terrace may be traced. From the flats of Cargen the land slopes gradually up to the 50-foot beach, which stretches from Cargenholm northwards to Park near Maxwelltown.

Along a great part of the coast there is a constant loss of land from the action of the sea. This loss is greatest where the sea-board is composed of boulder clay and other deposits, and the erosion is most rapid during severe storms blowing inshore. The material removed is not wholly lost ; some of it is carried inwards by the flood-tide and laid down as sediment on the fore-shore. Thus there is a twofold process continually at work : here and there the sea is gaining upon the land ; here and there land is being reclaimed from the sea. The shores of Loch Ryan have suffered considerably within the last hundred years. The Scar Ridge at one time extended about half a mile into the sea and cattle used to graze on it. So too on the western shore of Luce Bay, between Sandhead and Drummore, the sea has at several points gained upon the land ; while at the same time there has been an increase of the sandy foreshore at the head of the Bay. The estuary of the Cree shows both loss and gain. In some places many acres have been lost ; in others extensive reclamation has taken place, much of what was at one time soft marsh or sand being now grazing links. In Auchencairn Bay, a strip of merse-land on both sides and much land at the head of the Bay have been washed away within the last fifty years. The coast at the head of Orchardton Bay is specially subject to erosion. Along the low sandy shores of Kirkbean there are belts of links which have been slowly wrested from the sea.

All round our shores, wherever navigation is dangerous, are built lighthouses for the guidance of mariners. On Cairn Ryan Point, on the eastern shore of the loch, is a lighthouse showing a fixed light, visible twelve miles. On the east pier of Stranraer is another fixed white light, and on the west pier a fixed red light. Corsewall light is familiar to all who cross the North Channel. Its gleams of white and red light, visible sixteen miles, increase to intense brilliance and gradually fade away into darkness. From the top of the lighthouse on a clear day Ailsa Craig is conspicuous on the north, with the hills of Arran beyond ; Argyll and Ireland lie to the west ; while eastwards the eye sweeps the coast of Ayrshire from the Galloway Burn to beyond Ardrossan. Near Portpatrick is Killantringan light, which with its flash and eclipse may be seen for nearly twenty miles. At the extreme end of the headland, close to the edge of a cliff 210 feet high stands the Mull of Galloway lighthouse, with an occulting light visible twenty-five miles. Here the view is magnificent. From the Dumfriesshire heights in the north-east the eye circles by Kirkcudbrightshire and Ayrshire over Kintyre to the Paps of

Jura in the north-west ; twenty miles to the south the

Mull of Galloway

outline of the Isle of Man cuts the sky ; in the west are
seen the Mountains of Mourne ; while far away on the

eastern horizon loom the giant peaks of the Cumbrian Mountains. From Hestan Isle with its cave-riddled cliffs a white flash warns the sailor off the deadly stretch of Barnhourie Sands, and from Satterness a fixed white light repeats the tale.

10. Climate

By climate is meant the general tendency of a district towards mild or severe, average or extreme atmospheric pressure, temperature and moisture. Weather is the variation from time to time of all or any of these conditions. Thus climate is the mean of weather, and the two terms are symbols of different quantities of the same thing. Weather depends primarily on atmospheric pressure. This is measured by the barometer, which rises or falls as the weight increases or diminishes. In Britain in fine weather the barometer is usually above 30 inches, and is below this when there is rain or storm. For any given number of days on which the barometer stands at 30 inches, there are as many fine as rainy days.

The prevailing winds of Galloway are westerly and south-westerly. What is at once an effect and a demonstration of the cause is to be seen in trees grown in exposed situations. Their branches grow in an easterly direction. The south-west winds, by far the most common in winter, blowing from lower and warmer latitudes across the Atlantic, are the dominant factor in the climate of Galloway. Laden with aqueous vapour with which it has become impregnated in its passage

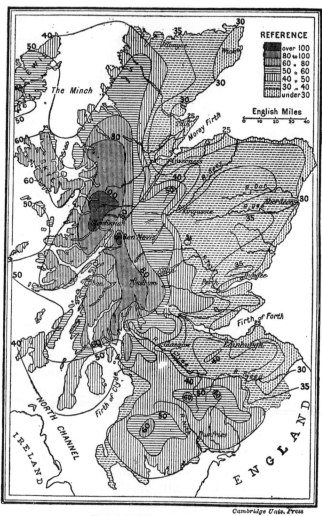

Rainfall Map of Scotland

(*By Andrew Watt, M.A.*)

over the ocean, the air on striking the land is forced
upwards wherever it meets with rising ground. Thus,
reaching a region of diminished pressure, it expands,
and has now a lessened capacity for holding water-
vapour, a portion of which is precipitated as rain. A
comparison of the rainfall map with the physical map
shows a marked correspondence between elevation and
rainfall ; the hilly regions are the wettest. The western
and lower part of Wigtownshire shows a yearly rainfall
of less than 40 inches ; the rest of the Shire with nearly
all the Stewartry is above that figure. Again nearly
all the northern portion of Kirkcudbrightshire is within
the 50-inch contour, and in the rugged mountainous
region in the north-west of the county there is a rainfall of over 60 inches. The influence of orographical
features upon amount and distribution of rainfall is well
shown by the following figures extracted, by permission,
from Mr Andrew Watt's *Mean Annual Rainfall of
Scotland*, 1871-1910.

Station.	Height above Sea.	No. of Years.	Mean Average Rainfall.
	Feet		Inches
Galloway House, Garlieston	20	30	39.73
Auchencairn . . .	50	30	47.41
Gatehouse (Cally) . .	120	40	49.43
Glenlee, New Galloway .	208	30	57.27
Carsphairn . . .	574	10	61.17
Carsphairn, Shiel . .	850	5	77.54

A rainfall record kept at twenty-one stations in Kirkcudbrightshire for periods varying from five to

Yews, Lochryan

forty years (ending 1910) shows the mean annual rainfall for that time to have been 53.68 inches. In Wigtownshire the stations are not so numerous. The mean for seven stations, ranging from a five- to a forty-year period, was 38.83 inches. This is about fifteen inches less than that of Kirkcudbrightshire, and is in accordance with the relief of the counties.

A temperature record kept at Cargen, Slogarie, Glenlee, Cally and Little Ross in Kirkcudbrightshire shows, for the forty years ending December 1895, a mean temperature of 32° F. for January and 48° F. for July— a mean annual range of 16° F. A similar record for Wigtownshire kept at Corsewall, Loch Ryan, Ardwell, Kirkcowan, and Mull of Galloway gives a mean January temperature of 40° F. with 57° F. for July—a mean annual range of 17° F. For Edinburgh the mean annual range is 21° F. and for London 26° F.

On the whole the climate of Galloway is favourable to health and longevity and to the agricultural pursuits upon which the province depends. Thanks to the south-west winds from the warm southern regions of the Atlantic, the winters are as a rule mild. Vegetation commences earlier in the spring and continues later in the fall than on the eastern coast of Scotland. Long continued frosts occur but rarely, and snow seldom lies long, at least in the lower districts. According to a work on the agriculture of Galloway published a hundred years ago, " It is generally calculated that in December and January the industrious farmer can plough on an average four days per week, and in November and February five." The statement holds good to-day.

E

11. People—Race, Dialect, Population

It is almost certain that the earliest inhabitants arrived in Britain when it was simply the west end of the Continent of Europe. They were small-boned, long-skulled and dark-haired, and they spoke a dialect of Iverian, a language whose descendant lives to-day on the lips of the Basques. After a time they were driven out or extirpated by invading Celtic tribes, who were long-boned, broad-skulled and fair-haired. To these the greater number of the place-names of Galloway are due, though the invaders would probably adopt and hand down to posterity at least some of the names of physical features as given by the conquered race. The name of the river Urr is practically identical with *ur*, the Basque word for water. In the first centuries A.D. the men of south-west Scotland were Brythonic—like the modern Welsh. Some of the best representatives of the Brythonic race, according to Dr Beddoe, are found among the tall hillmen of Galloway. But since most of the Celtic place-names in Galloway are not of Welsh but of Gaelic origin, it seems certain that there had been a large immigration of Gaelic speakers, perhaps from Ireland. Gaelic, indeed, continued to be spoken in Galloway to the end of the sixteenth century. The advent of Christianity introduced Latin words descriptive of Church offices and rites. " Sàgart, the priest (*sacerdos*) built himself a cill, a cell (L. *cella*) : so to this day Altaggart (*Allt Shaggairt*, the priest's stream) flows past the site of Kilfeather (*Cill Pheaduir*, Peter's Cell)."

Words of English origin passed in by way of Northumbria from the sixth century to the ninth. These in turn were supplemented by Scandinavian names brought by Norse marauders of the eighth to the tenth century. After the Norman Conquest a stream of Anglo-Normans poured northwards, bringing a further contribution to the language of Galloway, increased subsequently by English-speaking immigrants at the time of the Brucian settlement.

Such place-names as Bladnoch, Caitans, Rispain, Rotchell, Syllodioch date back to fable-shaded eras and their meaning is unknown. But the etymology of the great bulk of the place-names is fairly easy to make out. *Cnoc*, representing an isolated or precipitous hill, appears in over 220 place-names as prefix *Knock* ; and this is closely run by *drum* (*druim*), denoting the low glaciated ridge so frequently met with in the lower districts of Galloway. *Bar*, the top of anything, is a very common prefix. *Achadh*, arable land, is frequent as *auch*. Names of animals enter largely into the topography of both counties. Auchengower is the field of the goat ; Auchenlarie, the field of the mare ; Aucheness, the field of the horse ; Auchenshinnoch, the field of the foxes. *Pol, pal, pil, pul*, denote water, either flowing or at rest. *Darach*, the oak, and *beith*, the birch, give rise to scores of names ending in *darroch* and *bae*. Old English *burh*, *burg*, fortress, city, appears in Burrow Head ; *tun*, town, in Myreton, Broughton, Carleton ;

wic, village, in Rerwick, Senwick, Southwick ; *law*, hill, in Netherlaw, Wardlaw ; Norse *borg*, fort, occurs in Borgue, Borness ; *botl*, house, in Buittle ; *by*, dwelling, in Crosbie and Sorbie. Both syllables of Fairgirth are from Norse, and mean sheep-fold. So with Cogarth, enclosure for cattle, and Godgarth, enclosure for goats. *Knoits*, rocky hillocks, and *clints*, precipitous rocks, are characteristic of Galloway, as the Knoits of Bentudor and the Clints of Dromore.

To-day the vernacular of Galloway is a variety of Lowland Scots, and is most akin to that of south Ayrshire and west Dumfriesshire. Its written form, with its peculiar vocabulary and idiom, is very faithfully reflected in the novels of Crockett and in the works of several local poets.

Kirkcudbrightshire is ninth among the counties of Scotland in size : in population it is twenty-first. The actual figures from the census of 1911 are 38,367— 18,069 males and 20,298 females—for Kirkcudbrightshire, out of the total population of Scotland—4,759,445. This is 43 to the square mile, and gives about 14 acres to every man, woman and child in the county. Of the inhabitants above fourteen years 11,531 males and 4648 females were returned as engaged in one or other of the chief industries, while 1166 males and 10,390 females had no specified employment. Agriculture occupied 4870, and domestic service 2950. Connected with the building trades there were 875, including 354 joiners. There were 475 quarrymen and 365 metal workers. The textile industries employed 625, while drapers numbered 168 and tailors, dressmakers and milliners 1177. Nine hundred and fifty-six were engaged in the preparation and sale of provisions, 525 in railway service and road transit. The professional

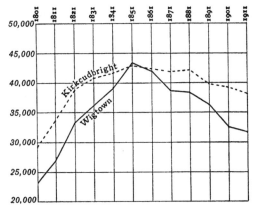

Diagram showing Rise and Fall of Population in Kirkcudbright and Wigtown since 1801

classes numbered 480, and 740 were engaged in Civil and Local Government Service.

Wigtownshire, ranking seventeenth in size, is twenty-third in population, the numbers being 15,078 males and 16,920 females—31,998 in all. This is 66 to the square miles, with 9 acres to each person. Above fourteen years of age there were 9338 males and 3672 females employed in one or other of the principal industries and services, while 1009 males and 8497 females had no specified employment. Agriculture engaged 5235 persons, and domestic service 1826. Including 241 joiners there were 565 connected with the building trade. Metal workers numbered 305 ; those engaged in textile industries 144. There were 155 drapers and 683 tailors, dressmakers and milliners. The preparation and sale of provisions occupied 998, while 533 found work in railway service and road transit. There were 325 members of the learned professions, and 541 attached to Civil and Local Government Service.

In both counties the alien element in 1911 was small. Kirkcudbrightshire had 85 foreigners, Wigtownshire 33.

12. Agriculture

The two counties are almost exclusively devoted to farming in one or other of its branches, sheep-rearing, dairying or mixed farming. Down to the middle of the eighteenth century the agriculture of Galloway was in the rude and barbarous condition common to Scotland. The farms were invariably over-cropped, and the methods of husbandry such that ten or twelve horses were required for the work now undertaken by two or three. Implements were often heavy and clumsy, always miserably inefficient. The soil was hopelessly impoverished by the practice of taking the same crop off it year after year as long as it would repay the seed and labour. The poor return of straw was inadequate for the needs of the always overstocked farm during the winter, and by spring the cattle were often so weak that they could not rise of themselves. Housing was poor beyond belief : wretched hovels built of stone and mud, thatched with fern and straw, unglazed holes for windows, no chimneys to give egress to the smoke, which found a tardy escape as best it could, were shared in common by the people and the cows of the farm, often without an intervening partition. But about 1750, with Mr Craik of Abigland (1703-98) and the school of farmers which subsequently formed themselves on his model, began the series of improvements in agriculture which have raised the Stewartry to its present high position among the counties of Scotland. Enclosing and draining the land, a regular system of fallowing, the use of calcareous manures as well as the liberal application of farmyard manures to fallows and fallow crops, the introduction of greatly improved implements, the establishment of a regular system of rotation of crops into which was introduced the use of sown grasses—these were the chief features of the new school.

Proprietors in Wigtownshire were no less eager to encourage and assist their tenants in the improvement and management of their farms, and this produced little short of a revolution in agriculture.

The arable part of Kirkcudbrightshire is found chiefly in the parishes which fringe the coast, in the eastern slope of the country, along the valleys of the Urr, Dee, Ken, and Fleet, and in the table-lands between these valleys. In Wigtownshire the line of railway from Newton Stewart to Glenluce may be taken approximately as the boundary between the high and low grounds, the cultivated area lying to the south of this line.   Arable farms run from 60 to 600 acres and are rented from £80 to £700 a year, few exceeding £1000. Hill or stock farms are, on the whole, much larger, few being rented under £200, while several exceed £1000. The rotation of crops is almost uniform.   The five-year course is—oats ; green crop ; oats (in Wigtownshire, barley or wheat) sown out with grasses and clover seeds ; hay, cut green, or seeded or pasture ; pasture.   The six-year course is the same with the addition of another year in pasture.   Wheat was extensively grown from 1815 to 1830.   As late as 1855 wheat in Wigtownshire covered 7343 acres ; in 1913 it covered only 71.

Galloway is earlier than most of Scotland.   Sowing of oats begins about the third week of March, and finishes as a rule by the middle of April.   Harvest, begun by the 12th or 15th of August, is finished in from three to five weeks, though some districts are two or three weeks later.

In 1913 of the 575,832 acres in the Stewartry, 92,458 were of arable land, 96,670 of permanent grass and 343,500 of mountain and heath land used for grazing. 25,293 acres under oats yielded an average per acre of 30.08 bushels ;   1709 acres of potatoes, 6.04 tons ; 11,166 acres of turnips and swedes, 16.17 tons ;   9465 acres of hay grown from rye-grass, 30.02 cwts. ;   12,670 acres of hay from permanent grass, 28.34 cwt.   In the same year the figures for Wigtownshire were : total area, 311,984 acres ;   arable land, 110,722 acres ;   per-manent grass, 8753, and mountain and heath-land used for grazing, 107,814. The 30,535 acres under oats gave a return of 36.79 bushels per acre; 1234 acres of potatoes, 7.5 tons; 14,167 acres of turnips and swedes, 16.5 tons; 4950 acres of hay grown from rye-grass produced 41.05 cwts., and 4731 acres of hay from permanent grass, 36.24 cwts.

Galloway cattle form one of the oldest and most characteristic of British breeds. They are essentially a beef-producing class. They are polled, and a coat of shaggy or curled black hair with an under coat of fine short wool fits them for the moist climate of the district. The picturesque Belted Galloways form one of the most valuable strains of this ancient breed. They are described as "exceptionally thick, blocky, nice-haired animals, and so hardy that they can winter and calve outside and ail nothing." One of the most interesting herds of Belties in Galloway belongs to Mr G. G. B. Sproat, Gatehouse-of-Fleet, the foundation of which was laid by his father in the Glenkens, early in last century. A two-year-old bull belonging to this herd scaled 15 cwts. in store order. A fine dairy of pure-bred Belted Galloways is owned by Mr James Brown of Knockbrex. A large and important branch of farming in Galloway is the rearing of polled store cattle for the markets of the south. These are bought for the most part as two-year-olds and are sent direct south to "gentlemen's grazings," the blue-grey shorthorn Galloway cross being a particular favourite in England. On dairying farms the stock used consists entirely of the Ayrshire breed of cattle.

In Kirkcudbrightshire few dairies have under 30 cows and few more than 70 or 80 : in Wigtownshire the numbers range from 20 to 350. The produce of the dairies is sold in various forms. Some send milk into

Belted Galloway Cattle

(Part of a fine herd belonging to James Brown, Esq., of Knockbrex, Borgue)

the surrounding towns. Many send their milk to one or other of the creameries, while many make it into cheese on the Cheddar system. The returns vary with the nature and amount of the food. In the southern half of the Rhinns, that Goshen for cheese-making, a dairy of 80 cows has averaged 19 stones of cheese per cow for six months. But over all 17½ stones per cow may be taken as the figure for Galloway dairies. This represents about 3400 pints of milk, or 1 lb. of cheese per gallon of milk.

A great many of the dairies are managed on either the bowing or the kaneing system. The farmer provides and keeps up the cows, buildings and dairy utensils, allows a certain area of pasture, a fixed quantity of roots and artificial food, with hay and straw *ad lib*. The bower pays his rent in money from £10 to £15 per cow, and does all the labour connected with the dairy, and receives all the produce in calves, cheese and pigs fed on the whey. The kaner pays rent in kind, about 19 or 20 stones (of 24 lbs.) of cheese per cow. The rent varies according to the quality of the pasture, and the amount and kind of the roots and artificial food supplied by the farmer.

The United Creameries, Ltd. has its headquarters at Dunragit, six miles from Stranraer, with branch factories at Sorbie, Wigtownshire, and Tarff, Kircudbrightshire. All milk is weighed and sampled on delivery and paid for on the basis of the butterfat contained. This is extracted by separators, the cream and the separated milk being delivered in different directions. Part of the cream is chilled to 35° or 40° F. and then put up in jars or cans for sale as required. The remainder, by far the larger part, is made into butter. The latest type of churn—a combined churn and butter-worker— is in use, in which not only is the cream churned into butter, but the butter is worked ready for packing. In the whole of the process neither cream nor butter is touched with the hand, the utmost cleanliness in manufacture being thus attained. The buttermilk with a very large proportion of the skim-milk is used for pig-feeding. A regular stock of ·from 2500 to 3000 pigs is kept, and the fat pigs are killed every week. Perhaps the most important branch of the business is the manufacture of margarine, for which large and thoroughly equipped premises are established at Dunragit. The margarine plant is capable of handling about 50 tons per week.

The Wigtownshire Creamery Co. has its central creamery at Stranraer, with branches at Sandhead and Drummore, Wigtownshire, and at Ballymoney, Ireland, all equipped with the most modern machinery. The Company handles milk from 9000 to 10,000 cows during the year, and manufactures cheese, butter and cream. It also sterilizes a quantity of milk and cream to be put up in air-tight stoppered bottles. A creamery at Bladnoch (with a branch at Whithorn), belonging to the Scottish Co-operative Wholesale Society, produces butter, margarine and margarine cheese.

Of the 55,398 cattle in the Stewartry in 1913 cows and heifers numbered 19,166, the remainder being two years and under. The corresponding figures for the Shire were 56,800 and 26,883.

Cheviot and black-faced sheep are almost the only stocks bred or fed in Galloway, and by far the greater number are black-faced. The class used depends on the produce of the land : where there is plenty of grass, even though poor, the Cheviot is the more profitable ; on land producing chiefly heather the black-faced is preferable. While equally hardy, the two breeds differ in quality of wool and mutton, the Cheviot possessing the finer wool, the black-faced the finer mutton.

In 1913 the Stewartry had 163,754 breeding ewes, with 219,145 other sheep ; the Shire had 48,552 and 62,339.

Brood sows numbered in the Stewartry 763, with 9989 other pigs ; in the Shire 840 and 15,443.

The old race of Galloway horses—" Know we not Galloway nags ? " asks Ancient Pistol in Shakespeare, 2 *Henry IV.*—strong, rough-legged hardy cobs about 14¾ hands high, and much esteemed for pluck and endurance, is now extinct. The whole attention of breeders has been turned to Clydesdales, the Scottish type of agricultural horse. Many of the best Clydesdales have been bred in Galloway. Little attention is paid to the breeding of saddle and driving horses, which in 1913 numbered 1104 in the Stewartry and 810 in the Shire. Of horses used for agricultural purposes, including brood mares, there were 3309 in the former county and 3498 in the latter ; of unbroken horses 1520 and 1642 respectively.

13. Manufactures, Mines and Minerals

The manufactures of Galloway are few and unimportant. Attempts made at various places to establish seats of manufacture have not met with lasting success, and to-day existing works do little more than supply local needs.   About 1778 a large factory for cotton spinning was erected at Newton Stewart.   But by 1826 the scheme, which for a few years had worked well, proved a failure, and the factory ceased work.   Hand looms, which in 1818 numbered 311 and whose products found a ready market with the merchants of Glasgow, had fallen in 1828 to a third of that number, and in

Newton Stewart

a few years the industry dwindled to extinction.   In 1790 Gatehouse-of-Fleet had two cotton factories, which gave employment to upwards of 200 hands, with a yearly output of nearly a million and a half yards of cloth.   But distance from the centres of population and the want of facilities for transport added greatly to the price of both the raw material and the manufactured article.   About 1815 decline set in, and by the middle of the century the works had shut down.   Part of the buildings is now occupied by a bobbin mill, employing about twenty men and boys. To-day woollen and tweed mills at Maxwelltown, at Twynholm, at Newton Stewart and at Kirkcowan give employment to 625 men and women in Kirkcudbright-shire and 144 in Wigtownshire.

Galloway is not a mining country. Laborious and expensive searches for coal have met with no practical success, as what was found in Kirkbean was in too small quantity to pay the expense of working. Veins of iron occur at various places in the Stewartry. One, to the west of Auchencairn, was worked for some time but was abandoned owing to the small returns and to the distance from a supply of coal. There used to be a copper mine in operation at Enrick, near Gatehouse-of-Fleet, the ores of which, green carbonate of copper and sulphate of copper with iron pyrites, are said to have yielded a rich percentage of the metal. An attempt to re-work this mine, made a few years ago, has been unsuccessful financially. In Colvend a copper mine for a brief period yielded a fairly rich ore from a tolerably thick seam. A vein of copper pyrites was formerly worked at Waukmill, near Kirkcowan ; two veins of barytes occur at Barlocco, Auchencairn, and one at Tonderghie near Whithorn. Galena has been mined at Blackcraig near Newton Stewart, in the Wood of Cree, at the Cairnsmore Mines, at the Pibble Hill Mines east of Creetown, at Woodhead near Carsphairn, and at Knockibae near Glenluce. But all these mines have been abandoned owing to scarcity of the mineral and expense of working.

Limestone of excellent quality is obtainable at Kirk-bean. Certain dark clays occurring at Brickhouse near Carsethorn have been worked for brick making, and there are brick and tile works at Dalbeattie and near Carty. Building material is obtained from the large quarries of the Queensberry grit group near Glenluce ; from the quarries near Newton Stewart and Wigtown ; from the porphyrites and micro-granites of Tongland and Loch Dougan ; from the granite quarries of Cree-town and Dalbeattie ; and from the beds of greywacke at Portpatrick. Wherever beds of greywacke are met with they are used for road metal, and Dalbeattie granite is largely used for the manufacture of grano-lithic pavement. At Cairnryan a band of grey shales and flags is worked for roofing purposes. The chief mineral wealth of Kirkcudbrightshire is its granite, and the quarries of Creetown and Dalbeattie are widely known. The Mersey Dock Board owned and worked one of the Creetown quarries and of its granite most of the Liverpool Docks were built. For eighty years this quarry employed from 180 to 300 men and had an average yearly output of 10,000 tons, half of which went for dock building purposes and half for setts. Much of the granite is crushed for use in pavements, garden paths, and such like. In Dalbeattie Messrs Fraser & Young make a specialty of the crushed granite trade. Their mills crush the stone and run it into railway waggons alongside. At their mill at Old Lands Quarry on the Urr vessels are loaded directly from the machine. The Craignair quarries have sent granite all over the world.   Lighthouses at Ceylon, the lower portion of the Eddystone Lighthouse, part of the Thames Embankment and of the Liverpool Docks, and the Albert Bridge, Belfast, for which more than 40,000 cubic feet of wrought granite were provided, are constructed of its stone.   Banks in London and Liverpool, the Town Halls of Manchester and Birkenhead, insurance buildings in London, Liverpool and other cities owe their structural beauty to the Granite City of the South.

Other industries are bone works and flour mills at Dalbeattie ; iron foundries and implement works, motor and coach works, cabinet-making works at Castle Douglas ; mills and dye works at Maxwelltown.

The extraction of salt from sea-water by evaporation was formerly carried on at several places on the coast. Satterness, now Southerness, and Saltpans Bay remind us of this industry by their names.   But with the repeal of the salt tax and the production of finer and cheaper salt in the " 'wich " towns of England the industry disappeared.

14. Fisheries, Shipping and Trade

Little has been done to develop and conserve the fishing industry of Galloway.   Symson's words in his *Large Description*, written in 1684, might be used to-day : " our sea is better stored with good fish than our shoare is furnished with good fishers."   Fishing is, of course, carried on at a number of places in Kirkcudbrightshire ;

F

but only two " creeks " are recognised by the Fishery Board for Scotland—Kirkcudbright and Creetown. In Wigtownshire there are ten—Stranraer, Kirkcolm, Portpatrick, Port Logan, Drummore, Sandhead, Glenluce, Port William, Isle of Whithorn and Garlieston.

Pelagic fish, including herring, mackerel and sparlings —the last got principally in the Cree—are taken by nets. Such demersal fish as cod, haddocks, skate, plaice and flounders are taken by trawl, lines and nets. Salmon are caught in fixed or " stake " nets. The principal shell-fish obtained are lobsters and crabs caught in dome-shaped cages of net stretched over a strong frame ; oysters taken by the dredge ; shrimps in specially constructed nets ; and mussels and whelks picked from the rocks to which they are found clinging. Recently whelks have been much over-gathered ; and the same is true of other shell-fish. In the estuary of the Cree hundreds of acres of mud cover to-day what forty years ago produced huge quantities of mussels. In former years thirty smacks at a time might have been seen dredging oysters in Wigtown Bay : this too is a thing of the past.

In 1913 there were 8991 fishing boats in Scotland, manned by crews amounting to 38,262. Of these Wigtownshire supplied 127 boats, none over 30 feet of keel, and 193 fishermen ; Kirkcudbrightshire boats numbered 21, all under 30 feet of keel, manned by 29 men. The total quantity of sea-fish of all kinds (exclusive of shell-fish) landed within the year was 7,828,350 cwts. of the value of £3,997,717. Of this amount Wigtownshire contributed 45,723 cwts.—32,354 cwts. being herring, valued at £14,321—which realised £19,647, and Kirkcudbrightshire 178 cwts., valued at £300. The total value of the shell-fish landed was £72,354. Of the 1,316,100 oysters included in this return, 1,305,400 oysters dredged from the beds in Loch Ryan sold for £4757. The shell-fish returns for Kirkcudbrightshire amounted to £537.

Salmon frequent the Cree, the Dee, the Fleet, the Nith and the Urr. The numerous lochs, well stocked as a rule with trout, and in many cases with perch and pike, offer excellent sport to the angler, while the burns and lanes of both counties contain sea-trout, herling, river-trout, pike and perch.

We must not omit reference to the Solway Hatchery, situated at Kinharvie, two miles from the village of New Abbey. It is one of the oldest and largest hatcheries in the kingdom, and from it large quantities of ova and fish are yearly dispatched to all parts of the world.

Notwithstanding the favourable length of seaboard, the commerce of Galloway is inconsiderable. There are few good harbours.

Portpatrick owed its early importance to its proximity to Ireland. In addition to mails and passengers, there were landed on its pier in 1812 no fewer than 20,000 Irish cattle. In 1821 operations were begun for the construction of a harbour on a large scale. Over £500,000 was spent in erecting sea-walls, deepening basins, and otherwise attempting to make it a safe haven. But the experience of a few winters with their tremendous gales from the south-west was sufficient to show that in the contest between man and the elements victory was to lie with the latter. The harbour was found unsafe, the mail-route was transferred to Stran-

The Harbour, Stranraer

raer and Larne, and to-day the sole shipping of this costly harbour consists of a few fishing boats.

Stranraer has a large and commodious harbour situated at the head of Loch Ryan. The loch itself is almost land-locked, and, except in the case of a gale from the north, the anchorage is all that could be desired. The harbour consists of a breastwork and an east and west pier. From the east pier steamers carrying mails, passengers and goods sail for Larne (39 miles) once a day in winter, and twice a day in summer. There is also regular steam communication with Glasgow and Liverpool.

The harbour of Kirkcudbright is well sheltered, of considerable extent, and of easy approach. But pier accommodation is very small, and tidal conditions make it suitable for small vessels only. Dalbeattie is served by a harbour on the river Urr, called Dub o' Hass, some five miles from the Solway, and vessels of 150 tons burden can come up thus far. At Old Land Wharf vessels of 200 tons can be handled, while Palnackie can be taken by vessels of 300 tons. The nature of the Solway beach and the phenomena of its careering tides render navigation precarious, and limit it on the whole to vessels of comparatively small tonnage.

From about the middle of the eighteenth century to that of the nineteenth, the story of the commerce of Galloway is in the main one of increase. Thus Kirkcudbright, which in 1801 had 37 vessels on its register, with an aggregate of 1648 tons, had in 1846 54 vessels, totalling 2069 tons. Wigtown, which had 25 ships with a burden of 984 tons in 1801, had in 1845 an aggregate tonnage of 3892. Stranraer, with 44 vessels capable of carrying 1732 tons in 1801, had in 1868 a tonnage of 2969. But with the introduction of railway facilities there came a sharp decline of sea-borne commerce. In 1913 Stranraer had only 8 vessels, aggregating 1886 tons ; Wigtown had 6, with a cargo capacity of 356 tons ; Kirkcudbright is now a creek under Dumfries, and the combined returns of port and sub-port showed for the same year a register of 14 vessels, with a tonnage of

770.   The trade of these ports is largely coastwise, and for the most part with towns on the west of England and Scotland, and east of Ireland.   It consists mainly of the import of coal, lime, and manures, and the export of agricultural produce.   The values of the imports from foreign countries for the year named were :—Dumfries (including Kirkcudbright), manures of all kinds £4519, oil-seed cake £852, all other articles £771 ;   Wigtown (including Garlieston, Port William, and Isle of Whithorn), manures of all kinds £1320, all other articles £405 ; Stranraer, manures of all kinds £1639, sawn wood and timber £1091, all other articles £1633.

OTHER STATISTICS OF GALLOWAY SEA-TRADE

VESSELS ENGAGED IN GENERAL COASTING TRADE IN 1913

	Entered.	Cargo in Tons.	Left.	Cargo in Tons.
Stranraer .	763	285,925	764	286,400
Wigtown .	213	16,468	212	16,674
Kirkcudbright	336	21,020	358	22,845

VESSELS ENGAGED IN FOREIGN TRADE

	Entered.	Cargo in Tons.	Left.	Cargo in Tons.
Stranraer .	5	2138	1	140
Wigtown .	3	399	Nil.	Nil.
Kirkcudbright	10	1106	Nil.	Nil.

15 History

Before the Roman general, Agricola, invaded North
Britain in A.D. 80, our knowledge of the history of the
country is scanty and untrustworthy.   In the course of
Agricola's campaigns, the Romans were in the south-
west corner in 82, looking out upon Ireland.   The district
afterwards to be known as Galloway, became, nominally
at least, part of the Roman Empire.   But the absence
of the remains of Roman camps and stations, and of any
Roman road west of the Nith, and the infrequency of
articles of Roman manufacture, show that the Roman
occupation was never very thorough, and was at most of
interrupted duration.

Towards the end of the fourth century, the first
Christian missionary (so says Bede, following tradition),
arrived in North Britain.   The most indubitable part
of the tradition is that St Ninian, landing at the
Isle of Whithorn, built a church of stone, which the
Latin writers knew as *Candida Casa*, in Old English
Hwitærn—both names meaning " White House."   St
Ninian converted the Southern Picts, as the men of the
region came later to be called ; but the new faith was
submerged in the old paganism, when early in the fifth
century the Roman power vanished in Britain.   Who
these Picts of Galloway exactly were is obscure.
Ptolemy had designated the inhabitants of the south-
west Novantae, and afterwards we hear of the district
as the home of the Attacotti.   These were distinct from

the Scots, and were absurdly credited with being cannibals.

Early in the seventh century Galloway fell into the hands of the kings of Northumbria, under whom the native chief ruled.   Anglians from Northumbria over-ran the district in considerable numbers, yet without effecting any great change in the district either in civil polity or in knowledge and practice of the arts.   Towards the close of the eighth century Northumbria was faced with the grim fury of the Northmen, and its suzerainty over Galloway had to be given up.   Galloway then sub-mitted to the sway of the Northmen, till freed from their domination by Malcolm Canmore about the middle of the eleventh century.   In 1124, on the accession of David I, Galloway became merged in Scotland.

When David interfered in the Civil War in England, the men of Galloway were prominent for their fierceness and their cruelty to the conquered.   In 1138, at the Battle of the Standard, they turned a probable victory into a defeat.   Their leaders claimed an ancient privilege of forming the van of the Scottish host, and though David knew the risk of exposing undisciplined troops, with no defensive armour, to the mail-clad Norman knights, he had to concede the claim.   All that stubborn courage could do, the Picts of Galloway did ; but the English arrows shot them down, and the Normans remained unbroken.   After two hours of grim conflict, the Galwegians lost their last chief, and on the cry that the king was killed, they turned in flight.   Then followed a general scattering of the Scots, though Prince

Henry's knights were winning in another part of the field. Only David's reserves prevented the English pursuit from annihilating the Scots.

In Malcolm IV's reign, Galloway rebelled, and was again subdued, only to break away when William the Lyon was taken prisoner in England. For eleven years Gilbert and his son were practically independent rulers. But after Gilbert's death in 1185, Roland, son of Uchtred, who had been murdered by his brother Gilbert, regained the lordship. By residence at the Scottish Court, and by marriage with de Moreville's daughter, Roland had become a Scoto-Norman, and was on friendly terms with King William. But even so, the men of Galloway, in the next century, more than once displayed their invincible love of independence, and their detestation of Norman ways.

When the Maid of Norway died, 1290, one half of the lordship of Galloway belonged to John Balliol, while a third of the remainder was owned by Alexander Comyn. In the war of succession which ensued, Galloway followed the banners of its lords and suffered accordingly. In 1300 Edward I overran Galloway as far as the Fleet, and reduced the Stewartry to subjection. It suffered again at the hands of Robert the Bruce, who invaded it because the inhabitants refused to follow his standard ; and the struggles of Edward Balliol to regain his father's throne once more plunged it into the horrors of war.

About 1370 Galloway came into the hands of the House of Douglas, and from then to 1455 the history of Galloway is a story of ravage and oppressive tyranny by the turbulent and ambitious family of Threave. On the fall of the Douglases the lordship of Galloway, with the earldom of Wigtown, passed to the Crown. Intestine strife, the consequence of frequent quarrels between petty chiefs, brings the history of the Province down to 1513, when many Galloway men of note fell beside their king on " Flodden's fatal field."

The doctrines of the Reformation were warmly espoused in Galloway. The attempts of the Stuart kings to establish Prelacy were resisted by none more strenuously than by the Westland Whigs. The Pentland Rising of 1666, the prelude to the " Killing Time," had its origin in Dalry, in the north of the Stewartry. At the battle of Bothwell Bridge (1679) a band of Galloway men in the Covenanting Army gallantly held the bridge against the Royalists till their ammunition was exhausted and they were ordered to retire. In the last years of Charles II's reign, and throughout James II's, the lonely moors and hillsides of Galloway were scoured by dragoons in search of Covenanters. Many a grave testifies to the steadfastness of the wild Westland Whigs, whom the troopers of Claverhouse and Grierson might kill but could not subdue. In May 1685 occurred the terrible drowning of the Wigtown Martyrs, Margaret M'Lauchlan, aged 63, and Margaret Wilson, aged 18, who were " by unjust law sentenced to die . . . and tyed to a stake within the Flood for adherence to Scotland's Reformation Covenants, National and Solemn League."

The Revolution settlement of 1689 was accepted quietly in Galloway. When the cry " The Auld Stuarts

back again " rang through Scotland in 1715, only two
Galloway gentlemen mounted the White Cockade—
Hamilton of Baldoon and Gordon of Earlston—and as
little interest was taken in the " 'Forty-five." In 1724
Galloway was thrown into confusion by the action of

Martyrs' Graves, Wigtown

the Levellers and Haughers, secret societies formed
against the Parking Lairds, who were endeavouring to
improve their system of husbandry by the erection of
march dykes and fences. But with the exception of
this episode the history of Galloway for the last two
hundred years has been one with that of the rest of
Scotland—material progress and general advancement,
social, educational and political.

16. Antiquities

In regard to early civilisations it is usual to speak of three epochs, the Stone, the Bronze, and the Iron. In the first, stone was the material used for those tools and weapons which, in a later and higher degree of culture, were made of metals. It is questionable whether palaeolithic man ever reached Scotland, but of the presence of neolithic man the evidence is ample. His weapons were of fine form, often highly polished, made of other stones than the flint of his palaeolithic predecessor, and are found associated with existing fauna.

In Galloway, cairns and hut circles, cliff forts and hill forts, mote-hills and doons indicate his distribution, and mark his activities. The stone circles and rock-sculpturings met with are referable probably to the bronze period. Cairns are classified as chambered or cisted. Of the former, which had within them a burial chamber capable of being used for repeated interments, there are eleven in Kirkcudbrightshire. Those which have long chambers lie in the valley of the Cree, east of which none is to be found south of Carsphairn. In Wigtownshire there are four, three in New Luce and one in Old Luce. Of cairns with round chambers there are four in the Stewartry, and three in Wigtownshire. Cisted cairns, containing a stone coffin intended for a single act of burial, are more numerous, and are widely distributed.

There are thirteen stone circles in the Stewartry,

three of which surround a central boulder. The only stone circle in Wigtownshire, with the outer ring of stones (19 in number) complete, is at Torhouskie near Wigtown. It is popularly believed to be the burial place of King Galdus. A stone at Laicht near Cairnryan, known as the Taxing Stone, is said to mark the tomb of Alpin, King of Scots, who was slain in Glenapp, 741 A.D. When circles are found in proximity to a cairn, they appear to have formed part of an original plan. A notable instance is the group of associated remains at Cauldside, Anwoth.

In Kirkcudbrightshire the area in which rock sculptures are found is restricted, but within this they are in considerable numbers. One group is found between the Cree and the Fleet, and another eastward from the estuary of the Dee to an imaginary line running north and south through Dundrennan. There is also a small group on the west side of Kirkcudbright. The greater number lie near the coast. In Wigtownshire such sculpturings have been recorded in ten places, most of these being in the Machers. The remains of ancient defensive constructions are very numerous in the Province. The Deil's Dyke was a rampart raised by the Galloway Picts as a defence against their neighbours to the north, the Brigantes of Strathclyde. According to Train " it commences at the farm of Beoch, and extends through the farms of Braid, Auchenvane, Kirnearven, and Kilfedder ; passes the north end of Loch Maberry, along Glenvernoch, and in Knockville runs into the Loch of Cree, to continue through Kirkcudbrightshire and Dumfriesshire as far as Hightae Flow in Loch Maben parish."

Not infrequently advantage has been taken of natural topographical situations, such as cliffs or promontories, or hills. Thus at Kemp's Wark, Larbrax, may be noted "the adaptation of its defensive lines to suit the altering requirements of the position as they pass from the narrow level of the front to the steeply sloping flank where they give place to a terrace." As a rule they are earth works, and consist of a single rampart and trench. But Borness Batteries, Borgue, is defended by two trenches and three ramparts, and the Doon, Twynholm, has double fosse and ramparts. The fort at Castle Hill Point, Colvend and Southwick, has for its main line of defence a stone wall some ten feet thick. Three hill forts in the Stewartry have been more or less vitrified, and one in the sister county.

Of mote hills, flat-topped mounds of earth and stone, in part natural, though sometimes wholly artificial, there are eleven in Wigtownshire and twenty-six in the Stewartry. The typical form is a truncated cone with an average height of 20 to 30 feet, surrounded at its base by a ditch. But the shape varies : at Boreland and Drummore, the mound is oval ; at Skaith it is almost a square. The most important in the Stewartry is the Mote of Urr. A simple truncated cone, it rises to a height of 33 feet, with a level top, 91 feet by 76, and comprises citadel, trenches, and base court on an extensive and well-preserved scale. The largest and best preserved in the Shire is the Mote of Innermessan. A

Mote of Urr

perfect circle, it has a circumference at the base of 336 feet, while from foundation to top it measures 78 feet.

Sculptured Stones, Kirkmadrine

It is generally believed that motes were used as courts of justice and places of public assembly, and in some places they are still known as court-hills. A curious broch-like structure at Castle Haven, Borgue, has had its details laid bare by excavation, and its construction

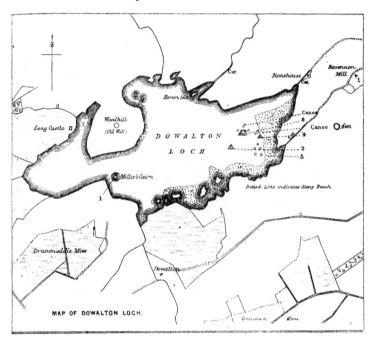

MAP OF DOWALTON LOCH.

restored. Of the many caves with which the shores are pierced, none is more deserving of notice than Borness Cave. It is situated at the head of an inlet below precipitous cliffs, about 27 feet above present high-water level. Systematically explored in 1872, it yielded abundant evidence of human habitation. The finds

G

included charred vegetable remains, remains of animals, polishers, whetstones, needles of bone, and a small cup of Samian ware, probably of the first century.

Early sculptured stones are very numerous in Wigtown-shire. Two now in a porch of the Church of Kirk-madrine, where they have been placed within recent

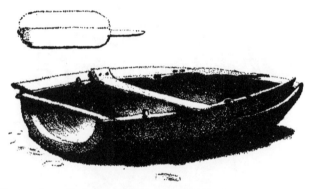

Canoe from Dowalton Loch, and Paddle from Ravenstone Moss, to the east of the Loch

years, are probably the earliest Christian monuments to the dead known in Scotland. Both bear the monogram of Christ within a circle, while a Latin inscription on one shows that it had originally indicated the last resting place of " two holy and pre-eminent priests." A long lost third inscribed stone was recently discovered acci-dentally. A sketch made a hundred years ago showed the inscription, " Initium et Finis, Alpha et Omega," with the Cross inside a ring, and the labarum of Con-

stantine on the top arm of the Cross.   A stone of high
antiquity is housed with other sculptured stones and old

Horned Mask of Bronze
(From Torrs, Kelton)

crosses in a crypt at Whithorn.   It probably marked
the position of a church dedicated to St Peter.

From the Glenluce Sands there have been recovered
" more objects of antiquity than from any area of similar
extent in Scotland."   The relics range from neolithic to

mediaeval times. The group of crannogs exposed by the drainage of Dowalton Loch in 1863, the first to be discovered of the unusually large number of these primitive dwellings in Wigtownshire, has done much to increase our knowledge of the manner of life and degree of civilisation of the ancient inhabitants of the country.

Bronze Bracelet
(Found near Plunton Castle)

The number of superimposed floors and the nature of the relics found embedded in the crannogs show that these lake-dwellings continued from neolithic times well into the Christian era.

Of miscellaneous objects of antiquity found in Kirk-cudbrightshire, the rarest is a small urn of the incense-cup shape, unearthed from an interment at Whinnyliggate. A horned mask of bronze was found at Torrs, Kelton; a bronze mirror in a bog, Balmaclellan; and a bronze bracelet near Plunton Castle. In Wigtownshire there fall to be noted a bronze axe, Glasserton ; a broad bronze dagger, near Stranraer ; and a gold penannular ornament, with ends terminating in cup-shaped discs, found on High Drummore, Kirkmaiden.

From time to time are unearthed flint-knives and arrowheads, discs, stone-hammers and axes, and finger-rings, which throw a dim and uncertain light on the life and customs of the by-gone races who roamed along the shores of the storm-bitten Solway, or trod the wind-swept moors of the interior.

17. Architecture—(a) Ecclesiastical

No trace now remains of the *Candida Casa*, the church built by St Ninian to the east of the Isle of Whithorn. But on its site stand roofless walls, part of a sacred edifice belonging, it is thought, to the thirteenth century, and probably a Chapel of Ease of the Priory of Whithorn. The Priory was founded in the twelfth century by Fergus, Lord of Galloway, who handed it over to Premonstratensian Monks. The church of the Priory became the cathedral of the diocese of Galloway, and remained so for 500 years. Here were deposited relics of the patron saint, and hither flocked crowds of pilgrims, of whom were kings and queens of Scotland, " For the dear grace to kiss St Ninian's bones." The nave of the Priory church, and a low fragment of a wall of the west tower are all that is left of the once stately pile. Recent excavations show that the total length of the church, from the west

tower to the Lady Chapel, was 250 feet.　In the south
wall is a splendid Norman doorway, dating back to the
foundation of the Priory, the rich carving on which has
bid defiance to the winds and rains of the centuries.
Of Wigtown Priory, a monastery of the Dominicans,
founded in 1267, and of Soulseat Monastery, no trace

St Ninian's Chapel

now exists.　By the end of the thirteenth century there
were in Scotland eight abbeys belonging to the Cistercian
Order, three of these being in Galloway—Glenluce, Dundrennan, and Sweetheart.　Glenluce Abbey, founded in
1190, was peopled with monks from Melrose.　Of the
church itself, Early English in style, there remains now
but the south transept gable, with eastern side chapels.
The cloister walls are fairly entire to the height of 16

feet, and the Decorated chapter house is well preserved, its arched roof supported by an octagonal pillar, 18 feet high.

The oldest religious house in the Stewartry is Dun-

Norman Arch, Whithorn Priory

drennan Abbey, founded 1142.   Its church was cruciform with a six-bayed nave, side aisles, transept and chancel, and central tower and spire, 200 feet high.   Built partly in the Transition Norman style, but belonging principally to the First Pointed, the chief portions extant are the

Dundrennan Abbey

north and south walls of the chancel, the east aisle of
the south transept, a few feet of the piers of the central
tower, and the doorway of the chapter house, flanked on
each side by a double window. Interesting monumental
stones are those known as the Abbot, the Cellarer, the
Nun, the Prior, and the Belted Knight.

In contrast to Dundrennan, the Old Abbey, Sweetheart is often called the New Abbey, because built 130
years later. New Abbey was founded by Devorgilla,
widow of the founder of Balliol College, Oxford. When
her husband died she had his heart embalmed and
placed in a casket, which she carried with her wherever
she went. She was buried near the high altar, the heart
of her husband being laid upon her breast. Hence the
romantic name, *Dulce Cor*, Sweet Heart. The abbey was
colonised by monks from Dundrennan, and was richly
endowed. The remains consist chiefly of the nave and
aisles of the conventual church. The mullions and
tracery of the western rose-window are fairly complete,
as also are the side windows of the choir, the clerestory,
and the upper windows of the north transept.

One other religious house falls to be mentioned, Lincluden Abbey, founded 1161 by Uchtred, son of Fergus,
Lord of Galloway, for nuns of the order of St Benedict.
Towards the close of the fourteenth century the nuns
became " insolent," and were expelled by Archibald the
Grim, who converted the foundation into an ecclesiastical
college. Of small extent, the Old College of Lincluden
is a very fine specimen of Gothic architecture. The
remains of the Collegiate church embrace the chancel,

south transept, south aisle and sacristy, and two vaulted chambers north of the sacristy.   In the middle wall of the choir is a magnificent tomb, canopied by a richly

Tomb of the Duchess of Touraine

ornamented semi-circular arch, in which was buried Margaret, wife of Archibald, fourth Earl of Douglas, who received the Dukedom of Touraine

From the Reformation to the opening of the nine-teenth century ecclesiastical architecture was practically dead in Scotland.   The eighteenth century churches

were " mean, incommodious, and comfortless." Since
then, however, a notable change has taken place, and
many a stately church has been reared in the country.

Glasserton Church

In Galloway we may take Glasserton as an example of
an eighteenth century building, repaired and ennobled
in the nineteenth century.

18. Architecture—(b) Military

Under the influence of Norman architecture, the old
single keeps of the Scottish landowners gave place to
stately piles of massive masonry, consisting of walled
enclosures, with towers of defence along the line of wall.
Of this, the Edwardian or First Period type of castle,

there are two examples in the Stewartry, and these are but fragmentary ruins—Castledykes, as the ancient castle of the Lords of Galloway is now called, at Kirkcudbright ; and Buittle Castle, about 1½ miles from Dalbeattie, which was also a stronghold of the Lords of Galloway, and which figured largely in the Wars of Independence, during the thirteenth and fourteenth centuries.

But the ravages of these wars impoverished the country and made buildings of such extent henceforward impossible. From the middle of the fourteenth century strongholds reverted to the simple keep, oblong in plan, with plain, massive walls, 8 to 10 feet thick. Gradually, however, they became more elaborate. They were once more built round a central courtyard, but the defensive features began to give way to domestic needs, while in some cases they were turned into ornaments.

The sixteenth century was a time of great activity in castle-building, the L plan being characteristic of the period. In this type a square wing, containing the wheel stair and small upper rooms, projects at right angles to the main building. Examples of this in the Stewartry are numerous, as the castles of Drumcoltran, Barholm, Carsluith, Kenmure, and Plunton. Of castles built on the Z plan Auchenskeoch is the only example in the county.

Threave Castle possesses unusual interest, because of its style of architecture, its association with many noteworthy incidents in Scottish history, and its ownership by the Douglases for nearly a century. It is built on an island of about 20 acres in extent, formed by two branches of the River Dee, about 2½ miles west of Castle Douglas.   It is protected by the main stream of the river on the west front ; on the other sides by a wall,

Threave Castle

5 feet thick, with round towers at the east angles and at the terminus of the south wall.   The tower at the south-east angle is still entire.   Its internal diameter is 9 feet, and it is surrounded by walls 4½ feet in thickness.   It is three stories in height, with three loopholes in each story.   A ditch, with a rampart outside the wall, enclosed an outer court, about 150 feet square, while

a gateway, defended by a drawbridge, but without a
portcullis, led through the east wall to the inner court,
and was opposite the entrance to the castle. The
keep measured 45 feet by 24 feet, within walls 8 feet
thick, which were pierced with windows on every side.
From the ground to the top of the ruined parapet on the
east side is fully 70 feet in height. The castle was built
by Archibald the Grim in the fourteenth century, and
is said to occupy the site of an earlier fortalice, of which
however, no traces now exist. Threave was the last
fortress to hold out for the Douglases, and the operations attending its reduction were superintended by
James II in person. The story of the siege, with the
part played by Mons Meg and her maker Brawny Kim is
firmly fixed in popular tradition, but does not bear close
scrutiny. After the castle became royal property, it
was entrusted to different powerful families in succession.
In 1526 it was vested in the Lords Maxwell as hereditary
keepers, who became Earls of Nithsdale and Stewards
of Kirkcudbright, and it remained in their hands till the
attainder of the Earl of Nithsdale in 1716.

Kirkcudbright Castle, standing on the left bank of
the Dee, belongs to the L type, with certain modifica-
tions. It is a strong, massive building, four stories in
height, its walls still almost entire. It was built in 1582
by Sir Thomas M'Lellan of Bombie, in whose family it
remained to the middle of the eighteenth century, when
it passed into the hands of Sir Robert Maxwell of
Orchardton: it is now the property of the St Mary's
Isle family.

Cardoness Castle, near Gatehouse, on the right bank of the Fleet, is a simple oblong, 43 by 22 feet.   It was

Cardoness Castle

built probably in the latter half of the fifteenth century. For centuries a seat of the powerful M'Cullochs, it is to-day owned by Sir Wm. Maxwell, Bart.   Three miles north of Gatehouse, also on the right bank of the Fleet, is Rusco Castle.   An oblong, 38 by 29 feet, it rises to a height of 50 feet, and is divided into three stories and

Rusco Castle

attics.   It dates from the close of the fifteenth century and was for long owned by the Gordons of Lochinvar.

Plunton Castle, Borgue, now in ruins, was built about the middle of the sixteenth century.   Cumstoun Castle,

Twynholm, also in ruins, is another sixteenth century building. The ruins of Wreaths Tower, Kirkbean, indicate the same period. Drumcoltran Castle, near Kirkgunzeon, is a sixteenth century erection. Midway between Gatehouse-of-Fleet and Creetown, and about a quarter mile from the coast are the ruins of Barholm

Hills Tower, Lochanhead

Castle. It is of the L type, and dates probably from the early years of the seventeenth century. With Carsluith Castle and several others, it claims to be the Ellangowan of *Guy Mannering*. Carsluith Castle stands on a promontory overlooking Wigtown Bay, about 3½ miles from Creetown. It is of L shape, with windows on the first floor. From time to time the building has been altered, the original part dating probably from the end of the fifteenth century. Hills Tower, Lochanhead, in Loch-

H

rutton parish, is an ancient building, with a later entrance lodge bearing date 1598.

The Round Tower of Orchardton, situated about 6

Round Tower of Orchardton

miles south-east of Castle Douglas, is the only one of its form in the province, and in some respects is said to be without parallel among the castles of Scotland. The tower is about 40 feet high, with an inside diameter of 15 feet, and consists of three stories. The second story appears to have been used as the principal apartment.   A circular piscina in this indicates its use at times as a private chapel. It dates probably from the latter half of the fifteenth century.

Of the Edwardian type of castle there appear to have been two in Wigtownshire, Cruggleton, of which there remains but a single arch, and Wigtown, of which there is now no trace.

The majority of castles in the county were erected between the fifteenth and seventeenth centuries ; and of these the Old Place of Mochrum, belonging successively to the Dunbars, the M'Dowalls, and the Bute family, is the most remarkable. It has been carefully restored, and is to-day an excellent reproduction of a late fifteenth or an early sixteenth century castle. Dunskey, Myrton, and Killaser belong to the same period ; Lochnaw and Craigcaffie date some fifty years later. Dunskey, a weather-beaten ruin, on an almost inaccessible headland overhanging the sea, with an immense ditch on the landward side, must have been impregnable. It was of the L type. Others of this type in the county are SorbieTower, Stranraer Castle, Myrton, Galdenoch, Castle Wigg, and Isle of Whithorn Castle. Castle Park, Glenluce, is the most complete example of the L castles built about the close of the century.

Craigcaffie, a fine old ruin, is another example of the square keep. Long the property of the Neilson family, it has formed part of the Stair estates since 1791. Stranraer Castle, built by Adair of Kinhilt, passed into the hands of the Kennedys, and thereafter to the Dalrymples of Stair. For a time it was used as a prison, and to-day most of it is occupied by merchants' stores. Castle Kennedy, built in the beginning of the seventeenth century, was destroyed by fire in 1716, and has never been rebuilt. A dormer window, with a beautiful head, is the

Castle Kennedy

only architectural ornament remaining to the ivy-clad ruins. On the shores of the White Loch of Myrton are the ruins of Myrton Castle, the keep of which was erected on a mote-hill. From the end of the eighteenth century it has belonged to the Maxwells of Monreith.

19. Architecture—(c) Domestic and Municipal

The venerable mansion of Kirkconnell is said to be one of the oldest inhabited houses in Scotland. It contains many interesting objects associated with the life of Mary Queen of Scots, and her descendants, James II and the Old Pretender.

Finely situated on a conspicuous knoll at the head of Loch Ken, stands Kenmure Castle, for centuries the principal seat of the Gordons of Lochinvar. The present building, which appears to have been built on the E plan, is said to occupy the site of one of the seats of the Lords of Galloway. Tradition says that John Balliol was born in the old fortalice, and that it became his favourite residence. In 1715 Viscount Kenmure, " the bravest lord that ever Galloway saw," threw in his lot with the Jacobites. Taken prisoner at Preston, he was executed, and the estates and title were forfeited. In 1824 these were restored to his grandson, but the title became extinct in 1847. The Castle is now a commodious and handsome residence. The stately beech-hedges and the avenue of fine lime trees are specially noteworthy. Among the family heirlooms are several old pictures and Jacobite relics.

Erected in 1763 but greatly altered in 1835, Cally House, with its spacious gardens and extensive policies, is situated amid picturesque surroundings. The columns of the portico are massive granite monoliths. The entrance hall is built of marble and contains some fine pieces of sculpture.

Among the many other mansion houses of the Stewartry may be mentioned Ardwall, Cardoness, Cassencarie, Goldielea, Kirkdale, Kirroughtree, Cairnsmore, Cumloden and Shambellie.

Lochnaw Castle has been in the possession of the

Lochnaw Castle

Agnew family for nearly six hundred years. It is delightfully situated on a green eminence surrounded by woods and overlooking a romantic loch. The line of buildings runs east and west and fronts the south. A central square tower five stories high, a portion of the " New " castle built in 1426 still remains and forms part of the modern building. The grounds contain many fine specimens of foreign coniferous trees.

Galloway House, built about the middle of the eighteenth century, stands in a beautifully timbered park. It consists of a central block with two projecting wings of the same height, its handsome front facing the west and overlooking Cruggleton Bay.

Lochinch Castle, the residence of the Earl of Stair,

Lochinch Castle

is built in the old Scottish Baronial style, exhibiting pepper-box turrets, rope mouldings, crow-stepped gables and carved projecting gargoyles. Its terraced gardens are of singular beauty. There is also a splendid pinetum, the principal feature of which is the great Araucaria Avenue, said to be the finest of its kind in the British Isles.

Other residences of note in Wigtownshire are the Old Place of Mochrum, Monreith House, Lochryan House, Logan House, Glasserton House, Physgill, Dunragit, Corsewall House, Penninghame House and Dunskey.

Dalbeattie has a Town Hall built of native granite, with a square tower and illuminated clock. The New Town Hall of Castle Douglas, built in 1862 to supersede

Old Place of Mochrum

the Old Town Hall of 1790, is of red free-stone. In Kirkcudbright there is a quaint Mercat Cross, dating from 1504. Behind it is the Old Tolbooth, an erection of Tudor times, with tower and spire built of stones taken from the ruins of Dundrennan Abbey.

The Court House in Wigtown is a handsome building of red and white freestone with a lofty clock tower. Flanking the Old Cross, a monolith about 10 feet high and 18 inches in diameter, stands the New Market Cross, an octagonal pillar about 20 feet high, rising from

a circular flight of steps.   The Old Cross is a fine specimen of the pillar crosses characteristic of many Scottish
burghs.   In Stranraer the New Town Hall, built of

The Tolbooth, Kirkcudbright

red and white freestone, owes its architectural effect
to its lantern spire and crow-stepped gables. The
Macmillan Hall in Newton-Stewart is the largest public
hall in the county, and houses the municipal offices of
the burgh.

Old and New Market Crosses, Wigtown

20. Communications

Prior to 1780 there was scarcely a road in the two counties worthy of the name. The original trackways had been largely due to horsemen, whose anxiety to avoid bogs and morasses had led them to beat out paths over hills of, in many cases, very steep gradient. These roads were kept in supposed repair by statute labour, parishioners being bound to give six days' work every year upon the parish roads. The old " military " road from Dumfries to Portpatrick (so-called because soldiers were employed in its construction) followed the original tracks and was carried from height to height, with no ostensible object. Fragments of this road may yet be made out as far west as Glenluce, where all further trace ceases.

About 1780 Parliament imposed an assessment for making and maintaining roads in Galloway. Roads subsequently constructed were upon more approved principles, and repairs were more systematically effected. Traffic was kept as low down as possible, since it was found to be easier and cheaper to carry a road round a hill than over it.

Many of the old moor roads owe their origin to the ling-tow-men or smugglers. One of these ran from Portpatrick to Clydesdale by way of Loch Inch, New Luce, The House of the Hill and the Nick of Balloch. From The House of the Hill a road ran to Edinburgh by Glentrool and another to Ayrshire by the Nick of

Balloch. The routes from several landing-places converged at Kirkcowan, which thus formed a convenient halting-place on the way to Glasgow by Minnigaff, northwards by Loch Trool, Loch Enoch, Loch Doon and Dalmellington ; and to Edinburgh by Curriedon, Moniaive and Penpont, through the Dalveen Pass and past Elvanfoot. We must remember that the Galloway coast afforded unrivalled opportunities for smuggling, and in the eighteenth century and the first part of the nineteenth the "free-traders" plied a busy trade in brandy, silks and lace from the Isle of Man. Scott in a note to *Guy Mannering* mentions the statement of a smuggler "that he had frequently seen upwards of two hundred Lingtow-men assemble at one time, and go off into the interior of the country, fully laden with contraband goods."

Galloway now possesses excellent roads. We begin with Wigtownshire. At Challoch, 2½ miles from Newton Stewart, the road to Ayrshire divides. The right fork goes through the valley of the Cree as far as Bargrennan Church and thence across the north-west of Kirkcudbrightshire to Straiton. The left fork passes Glassoch and the Snap, crosses Fyntalloch Moor, and, leaving the county by the isthmus between Loch Maberry and Loch Dornal, makes for Barrhill. The road from Newton Stewart to Portpatrick follows in the main the railway line. At Glenluce it is joined by a road which left the road to Barrhill at Glassoch, and by one which has come from Girvan down the valley of the Luce. A road from Glenluce strikes Luce Bay at Auchenmalg and follows the coast as far as Port William. Thereafter it passes through Monreith and goes by way of Glasserton to Isle of Whithorn. About half a mile from this village it connects with a road which comes from Newton Stewart through Wigtown and Kirkinner to Sorbie. Soon after passing Glenluce the main road to Portpatrick forks at West and East Challoch, one branch crossing the Rhinns to its terminus on the North Channel, the other going by way of Stranraer along the eastern shore of Loch Ryan to the Galloway Burn, where it enters Ayrshire. Good roads also connect Stranraer with Corsewall Point and the Mull of Galloway.

Wigtownshire has three lines of railways. The Portpatrick and Wigtownshire Joint-Railway, branching off the G. and S.W. system at Castle Douglas, traverses the county by Newton Stewart and Glenluce to Stranraer and Portpatrick. The Wigtownshire Railway runs from Newton Stewart, by Wigtown and Garlieston, to Whithorn. The Girvan and Portpatrick Railway enters the county in the north of New Luce parish, and, following closely the valley of the main Water of Luce, joins the Portpatrick Railway at East Challoch near Dunragit.

Let us now turn to Kirkcudbrightshire. An excellent road leads from Maxwelltown to Newton Stewart by Crocketford. Skirting Auchenreoch Loch, it goes through Springburn on its way to Castle Douglas. It passes through Bridge of Dee, Ringford and Twynholm to strike the coast near Gatehouse-of-Fleet. From this point on to Creetown is often spoken of as the most beautiful shore-drive in Britain, there being in Carlyle's opinion only one to equal it—the drive back. Crossing the railway near Palnure Station, the road enters Newton Stewart through its picturesque suburb, Cree-bridge. From Maxwelltown the road to Dalbeattie follows in the main the railway, crossing at Kirkgunzeon Station from the south to the north side of the line. Starting once more from Maxwelltown, one may follow a good road south through New Abbey and Kirkbean and thence west to Rockcliffe. A branch connects this popular watering place with Dalbeattie. From Dalbeattie Kirkcudbright may be reached either by Castle Douglas, or by Palnackie, Auchencairn and Dundrennan. At Crocketford a branch from the main road makes for the north of the county through Corsock Bridge (where it is joined by one from Dalbeattie) past Balmaclellan to New Galloway. From Castle Douglas a road skirts the railway through Crossmichael to Parton. It leans upon the shore of Loch Ken for about a mile, and keeps within half a mile of the Loch till opposite Kenmure Castle. Near Dalbeattie it joins the road from Maxwelltown, and then, flanked by the grand hills guarding the Glenkens, it makes by way of Dalry and Carsphairn for Dalmellington and Ayrshire. From Kirkcudbright through Ringford and Laurieston a road which skirts the beautiful Woodhall Loch, crosses the railway at New Galloway Station. After hugging the western shore of Loch Ken for nearly three miles, it passes through New Galloway, and at Allangibbon

Bridge connects with the road coming through Dalry on its way north. A little frequented but highly picturesque hill road connects New Galloway with Newton Stewart. Many inferior roads and rough hill tracks cross the county and link up parishes and hamlets and farms with the more important centres.

From Dumfries the G. and S.W. Railway sends off a branch, which passes by Maxwelltown and Dalbeattie to Castle Douglas. It is continued to Creetown and Palnure, near which it enters Wigtownshire. From Castle Douglas the line to Kirkcudbright passes through Bridge of Dee and Tarff stations.

21. Administration and Divisions

When Galloway came under the dominion of the King of Scots, the inhabitants were allowed to retain their old laws, and this continued for a long time. These laws were to some extent modified by William the Lyon. When Archibald the Grim obtained the Stewartry and the Shire, he managed to secure the suppression of some of the old laws, but others remained till, by Act of Parliament in 1426, Galloway was brought under the general law of Scotland. But for long, indeed down to 1747, when heritable jurisdictions were abolished, the powers of both Steward and Sheriff were confused and overlapped by independent juris-dictions held by the great families and the great churchmen.

The custom of handing down responsible judicial

offices from father to son without respect to qualification for the position was perhaps vicious, but in actual practice it worked out not so badly. It was comparatively seldom that justice was perverted, seldom that decisions were partial, or that oppression was sustained. For long the Wild Scots of Galloway preferred " gentleman's law," the law of the heritable functionary to whom they instinctively yielded deference, to that of the more learned stipendiaries whose law not infrequently seemed at singular variance with native ideas of justice and equity.

In addition to jurisdictions of a baronial or feudal character there was and still is that of the burgh. There is the royal burgh, a corporate body erected to be holden of the Sovereign. The burgh of barony holds its charter from the feudal superior of the lands. Of more recent creation is the police burgh, a town or place of more than 700 inhabitants, made a corporation by Act of Parliament. There are six burghs in the Stewartry, Castle Douglas, Dalbeattie, Gatehouse, Kirkcudbright, Maxwelltown and New Galloway. The royal burghs are Kirkcudbright, since 1455, and New Galloway, since 1630. In the Shire, the royal burghs, in order of creation, are : Wigtown, 1457, Whithorn, 1511, Stranraer, 1617. Newton Stewart, originally a burgh of barony, is now a police burgh.

Burghs are managed by Town Councils. The Councillors regulate the trade of the burgh and the conduct of the inhabitants, and from their own number elect magistrates, who act as judges in the police courts.

County matters were formerly administered by Commissioners of Supply, but are now in the hands of the County Council. It levies rates for county purposes, it makes by-laws for the government of the county, it administers the Food and Drug Acts, and the Diseases of Animals Act, it maintains roads and bridges, it controls the police, it appoints officers of health, manages lunatic asylums and hospitals, and exercises general supervision over matters relating to public health.

The two chief authorities in a Scottish county are the Lord-Lieutenant, who is at the head of the magistracy and is the highest executive authority, and the Sheriff, who is the chief local judge of the county. The Sheriff is assisted by a Sheriff-substitute, or by Sheriffs-substitute. Kirkcudbrightshire has a Lord-Lieutenant, twenty-five Deputy-Lieutenants, and over 160 Justices of the Peace. Wigtownshire has a Lord-Lieutenant, thirteen Deputy-Lieutenants, and some eighty Justices of the Peace. Both counties have the same Sheriff-principal, each has a Sheriff-substitute. Kirkcudbrightshire has three Honorary Sheriffs-substitute, Wigtownshire five.

Kirkcudbrightshire and Wigtownshire are united for Parliamentary Representation into the Constituency of Galloway.

The parishes in Kirkcudbrightshire are : Anwoth, Balmaclellan, Balmaghie, Borgue, Buittle, Carsphairn, Colvend, Crossmichael, Dalry, Girthon, Irongray, Kells, Kelton, Kirkbean, Kirkcudbright, Kirkgunzeon, Kirk-

I

mabreck, Kirkpatrick-Durham, Lochrutton, Minnigaff, New Abbey, Parton, Rerrick, Terregles, Tongland, Troqueer, Tywnholm, Urr. The parishes of Carsphairn, Kells, Dalry and Balmaclellan are often spoken of collectively as " the Glenkens."

The Wigtownshire parishes are : Glasserton, Inch, Kirkcolm, Kirkcowan, Kirkinner, Kirkmaiden, Leswalt, Mochrum, New Luce, Old Luce or Glenluce, Penninghame, Portpatrick, Sorbie, Stoneykirk, Stranraer, Whithorn, Wigtown.

Till 1894 Parochial Boards looked after the affairs of the parishes, but were then superseded by Parish Councils. These administer the Poor Law, appoint registrars, provide burial grounds, and levy rates for education.

The Education Act of 1872 set up a new and homogeneous system of education in Scotland. School Boards were created in every parish and burgh in Scotland, and to them was entrusted the management of education within their bounds. Under the Munro Act of 1918 the School Board gives place to the Education Authority, and the parish as administrative unit of education to the county with its electoral divisions. The Act virtually recasts the whole of the Scottish educational system outside the Universities. Nursery schools may be instituted for children between the ages of two and five, and the age for leaving school has been raised to fifteen, with conditional exemption. Continuation classes, compulsory to the age of eighteen for those who are not receiving suitable instruction in other ways, are to give due attention to physical exercises, cultural subjects, and such vocational training as is suitable to the requirements of the locality ; and the pupils are to have the benefit of medical examination and supervision. Higher education up to Training College and University is to be made possible, by adequate financial assistance, for every child who can profit thereby. Thus extensively and intensively the Act is far-reaching, providing for the full educational development of the community, with equal opportunity for all.

22. Roll of Honour

The sons of Galloway have distinguished themselves in many walks in life. Brave men of action, learned jurists, pious and scholarly churchmen, philosophers, poets, novelists and artists have shed lustre on the Province which gave them birth.

Admiral Sir John Dalrymple Hay who was born in 1821 and died in his 91st year, could look back on a naval career of fifty years full of adventure and honour. Rear-Admiral Sir John Ross, a native of Inch, made several voyages of discovery in Arctic regions and published books and pamphlets on the results. Paul Jones, born at Arbigland in 1747 and known, till he transformed his name, as John Paul, was a famous seaman. When the American Colonies rebelled against Britain, he became head of their first naval force and made a descent on the Solway. He raided St Mary's

Isle, carrying off Lord Selkirk's plate, which he afterwards restored. Subsequently he served as rear-
admiral of the Russian Black Sea Fleet. Sir Andrew
Agnew, the last of the Hereditary Sheriffs of Galloway,

Sir John Ross

was in early life a skilful officer under the Duke of Marl-
borough and noted for deeds of great personal daring.
John Dalrymple, second Earl of Stair, distinguished
himself at Malplaquet and Ramilies. Sir William
Gordon of Earlston, an officer in the 17th Lancers, was
one of the " Noble Six Hundred."

Andrew Symson, who died in 1712, minister of Kirkinner for twenty years prior to the Revolution, though not a native, is closely identified with the Province by his *Large Description of Galloway*. Samuel Rutherford, covenanting hero and divine, was for nine years minister of Anwoth. John Macmillan, the founder of the Cameronian Church, was a native of Minnigaff.

Balsarroch

(At one time the property of the ancestors of Sir John Ross)

Dr Alexander Murray, a shepherd's son born at Dunkitterick, in the brief thirty-seven years of his life rose to be the most eminent linguist and Oriental scholar of his day. Dr Henry Duncan, a son of the manse of Lochrutton, minister of Ruthwell, was the founder of Savings Banks. Wm. Maxwell Hetherington, D.D., a native of Troqueer, church historian and poet, was Professor of Apologetics and Systematic Theology in the University of Glasgow. Alexander Raleigh, D.D., born in the parish of Buittle, was a prominent Congregational minister.

Professor Thomas Brown, who succeeded Dugald

Rev. Alexander Murray, D.D.

Stewart in the chair of Moral Philosophy, Edinburgh University, was born in Kirkmabreck manse. David Landsborough, the Gilbert White of Arran and the Cumbraes, was a native of Dalry. John Ramsay M'Culloch, born in Whithorn, in his day a noted writer on political economy, edited the *Scotsman*, 1818–1820.

Another journalist, William M'Dowall, published a valuable history of Dumfries as well as other works of a more or less antiquarian cast. The quaint *Gallovidian Encyclopedia* of John Mactaggart is a classic authority on Galloway customs and speech. With this work must be conjoined *The Seasons* by David Davidson, another Stewartry man. Will Nicholson, the author of the *Brownie of Blednoch*, is *the* Galloway poet. After him may be mentioned John Lowe, whose *Mary's Dream* was long a popular song in the district ; and Robert Kerr, best known by *My First Fee*, and *The Widow's ae Coo*. The Rev. William Mackenzie, a native of Kirkcudbright, wrote a laborious and minute *History of Galloway*. *The Literary History of Galloway* is from the pen of Dr Thomas Murray, a native of Girthon. In a series of Galloway stories embodying the spirit of the Province, Samuel Rutherford Crockett has fixed much of its folk-lore and legend, and a wealth of its old-world words and phrases. Several of the Trotter family, descendants of the famous " muir doctor " of Galloway, have won distinction as writers. His son Robert published tales founded upon local traditions. His daughter, Isabella, wrote memoirs of her father. Alexander Trotter, a grandson, was the author of *East Galloway Sketches*, and Robert de Bruce Trotter of two delightful volumes of *Galloway Gossip*.

The legal profession is represented by the first Viscount Stair, Lord President of the Court of Session, whose work, the *Institutions of the Law of Scotland*, is the greatest of the complete treatises on Scots Law ;

and by his son the first Earl of Stair, who succeeded " Bluidy Mackenzie " as Lord Advocate. Of recent years, Lord Ardwall was a Stewartry man in all but the accident of birth and early life. The same may be said of the great physicist, James Clerk Maxwell.

In art the Faed brothers have a reputation that is world-wide. John, the eldest, for many years was a noted miniature portrait painter. He was elected R.S.A. in 1851. Thomas, Royal Academician in 1864, excelled like his brother in subjects dealing with pathetic or sentimental incidents in humble Scottish life. A third brother, James, achieved high artistic success in line engraving.

23. The Chief Towns and Villages

(The figures in brackets after each name give the population in 1911, and those at the end of each section are references to pages in the text.)

A.—KIRKCUDBRIGHTSHIRE

Auchencairn (235), a village beautifully situated on bay of same name, about 10 miles east of Kirkcudbright, has good sea-bathing. Near it is Auchencairn House, with a fine collection of modern British paintings. (pp. 58, 63, 79, 126.)

Balmaclellan (pa. 559), a village in the N.E. of the county. Robert Paterson, Scott's " Old Mortality," lived here in 1768 ; and here his wife taught a small school for twenty years. (p. 126.)

Borgue (pa. 1023), a village 6 miles S.W. of Kirkcudbright. Near it is Earlston House. The parish has long been famous for its honey. (pp. 3, 68.)

Carsphairn (pa. 360), a village in extreme north of county, is a health resort. (pp. 6, 63, 79, 92, 126.)

Castle Douglas (3016), the commercial capital of the Stewartry, is a railway junction. It has a large well-equipped and highly successful Academy, has iron foundry, motor works, coach works, sawmills, cabinet-making works, aerated-water manufactories, a tannery and large grain stores. Castle Douglas is one of the most important market-towns in the south of Scotland, with busy sales of live stock every week, while hiring, horse and other fairs are held periodically. (pp. 15, 16, 33, 81, 109, 114, 120, 125, 126, 127, 128.)

Creebridge (366), a small village on Stewartry side of Bridge over the Cree at Newton Stewart. (p. 126.)

Creetown (873), a burgh of barony, seaport and fishing village, at head of Wigtown Bay, has large granite quarries. (pp. 23, 33, 42, 49, 50, 57, 79, 80, 82, 113, 126, 127.)

Dalbeattie (3357), the Granite City of the South, 14½ miles S.W. of Dumfries, has bone works, flour mills, dye works, brick and tile works, an iron forge, concrete works, wood-turning works, bobbin mill, saw mill, paper mill, and creamery. Its quarries, which employ several hundreds of men, yield very fine granite. (pp. 19, 33, 80, 81, 85, 108, 120, 126, 127, 128.)

Dalry (490), "The Clachan," "St John's Town," a village beautifully situated on the left bank of the River Ken, 16 miles N.W. of Castle Douglas, has good golfing. (pp. 6, 19, 30, 126, 127, 134.)

Dalry

Dundrennan (101), a village 5 miles S.S.E. of Kirkcudbright, charmingly situated in a narrow valley on the right bank of Abbey Burn, has ruins of a fine Abbey. (pp. 16, 93, 102, 103, 105, 126.)

Gatehouse-of-Fleet (1032), picturesquely situated on River Fleet about 9 miles from Kirkcudbright. A bobbin mill employs about twenty hands. Near it is Barlay Mill, the birthplace of the Faeds, the noted family of artists. (pp. 4, 16, 21, 25, 49, 58, 63, 73, 78, 111, 112, 113, 125, 128.)

Kirkcudbright (2205), the county town, royal and parliamentary burgh, on left bank of River Dee, 6 miles from the mouth of its estuary, has a very good museum, especially rich in flora and fauna of district. The Academy is a very successful secondary school and centre for junior students. (pp. 2, 3, 20, 30, 32, 33, 82, 85, 86, 93, 108, 120, 126, 127, 128, 135.)

Kirkpatrick-Durham (277), a village 5 miles N.N.E. of Castle Douglas.

Kippford (pa. 696), 4 miles south of Dalbeattie, the most important of the Colvend watering-places ; headquarters of Urr Yacht Club. Kippford does a large trade in mussels, and has good golfing. (p. 46.)

Maxwelltown (6200), formerly the " Brig En'," on the right bank of the Nith, directly opposite Dumfries, has tweed mills, hosiery manufactures, dyeworks, saw mills and nursery grounds. H.M. General Prison for Dumfries and Galloway is situated in the town. The Observatory Museum contains many relics of Burns, and a fine collection of minerals. There is also a camera obscura with regular suite of Claude Lorraine glasses. About a mile from Maxwelltown are the ruins of Lincluden College. (pp. 7, 34, 36, 58, 79, 81, 125, 126, 127, 128.)

New Abbey (178), a village 7 miles south of Dumfries. The ruins of Sweetheart Abbey are close to the village. Near it is the Solway Fishery, one of the largest hatcheries in the kingdom. (p. 83.)

New Galloway (352), a royal and parliamentary burgh at head of Loch Ken, has good golfing. Near it is Kenmure Castle. (pp. 6, 19, 33, 63, 126, 127, 128.)

Lincluden College

Palnackie (pa. 825), a village in Buittle, on right bank of Urr Water, 3¾ miles S.S.W. of Dalbeattie, has a good natural harbour ; and, till the introduction of the railway in 1861 diverted its trade, was the port of Castle Douglas. (pp. 19, 85, 126.)

Rhonehouse, a village 2½ miles from Castle Douglas, formerly noted for its fairs, one of which was the most important in the south of Scotland.

Rockcliffe (72), a hamlet in Colvend, 7 miles S.E. of Dalbeattie, an excellent watering-place. (pp. 6, 46, 126.)

Southerness Village, in Kirkbean parish (711), 10 miles S.E. of Dalbeattie, a favourite resort of sea-bathers and summer visitors. (pp. 43, 58, 81.)

Sweetheart Abbey

Twynholm, a village 3 miles N.N.W. of Kirkcudbright, has an old established woollen mill, where tweeds and blankets are manufactured. (pp. 3, 9, 79, 125.)

B.—WIGTOWNSHIRE

Bladnoch (pa. 1369), a village on river of same name, $1\frac{1}{4}$ miles from Wigtown, has a large distillery and a creamery. The Wigtown Martyrs were drowned in the river, 1685. (pp. 67, 76.)

Cairnryan (pa. 1860), formerly Macherie, a seaport village on eastern shore of Loch Ryan, has a good harbour. (pp. 56, 93.)

Drummore (401), a seaport village on the west side of Luce Bay, has a small harbour with good anchorage. In the immediate vicinity there is splendid bathing ground. (pp. 53, 58, 76, 82, 94.)

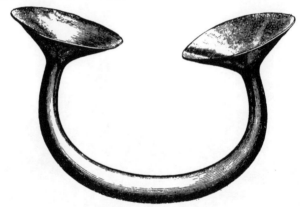

Gold Penannular Ornament
(Found on High Drummore)

Garlieston (482), a village about 9 miles S.E. of Wigtown, has boat building, fishing, chemical manufactures, grain mill, saw mill, and considerable export of whelks. (pp. 63, 82, 86, 125.)

Glenluce (774), a village 8 miles east of Stranraer. Two miles north are the ruins of Glenluce Abbey. (pp. 3, 53, 71, 79, 80, 82, 102, 115, 123, 124, 125.)

Isle of Whithorn (261), a seaport village 3¼ miles S.E. of Whithorn ; popular summer resort. (pp. 52, 82, 86, 87, 101, 125.)

Kirkcowan (pa. 1244), a village on left bank of Tarff Water, 6½ miles from Newton Stewart, has woollen mills. (pp. 33, 65, 79, 124.)

Kirkinner (pa. 1206), a village about 3 miles south of Wigtown, noted for its fine scenery.   (p. 125.)

Creamery, Drummore

New Luce (pa. 481), a village on left bank of Water of Luce.   " Prophet " Peden was minister of the parish for three years prior to his ejection in 1662.   (pp. 23, 92, 123.)

Newton Stewart (2063), finely situated on right bank of River Cree. A tannery, a brewery and tweed mills give employment to a number of hands.   Wool furnished from the surrounding country and purchased for the English markets has for long been a staple branch of its trade. The Douglas High School for girls and the Ewart Institute for boys are successful secondary schools. (pp. 23, 71, 78, 79, 80, 121, 124, 125, 126, 127, 128.)

Port Logan (pa. 1792), a fishing village at the head of Portnessock Bay, 14 miles from Stranraer : is the station of a lifeboat which serves the Bay of Luce and the Irish Channel. Near it is the Logan fish pond, constructed in 1800, into which the sea washes at every tide through a narrow crevice. It is visited annually by hundreds to see the tame fish, cod and saithe, which are kept in it. (pp. 36, 82.)

Port William (645), a seaport on east side of Luce Bay, 24 miles S.E. of Stranraer. (pp. 53, 82, 86, 125.)

Portpatrick (517), a village picturesquely situated amid fine cliffs on the west coast of the Rhinns, is a very popular holiday resort. The town owed its early importance to its nearness to Ireland. As far back as 1677 a boat carried mails twice a week to and from Donaghadee, 21 miles distant. (pp. 6, 9, 55, 59, 82, 83, 123, 124, 125.)

Sandhead (pa. 2279), a village on Luce Bay, 7 miles south of Stranraer, has considerable fishing. (pp. 53, 58, 76, 82.)

Sorbie (pa. 1354), a village 6¼ miles south of Wigtown. Two miles west was Dowalton Loch, now drained, famous for its crannogs. (pp. 68, 75, 125.)

Stoneykirk (pa. 2279), a village in the Rhinns, 6 miles S.S.E. of Stranraer. The name is derived from St Stephen, in Scots, Steenie. This word was by mistake regarded as coming from *stane*, Scots for *stone*.

Stranraer (6444), a royal and police burgh at the head of Loch Ryan, is the herring-fishing headquarters for boats on the Ballantrae banks and a centre and market for a large agricultural district, with cattle, horse and hiring fairs. It has flour mills, creameries, and a noted oyster fishery. Its castle was the residence of Claverhouse when

Creamery, Sandhead

Sheriff of Galloway. Stranraer is in direct communication by rail with Carlisle and Glasgow, and by sea with Larne and the north of Ireland. (pp. 6, 7, 28, 56, 59, 75, 76, 82, 84, 85, 121, 125, 128.)

Whithorn (1170), a royal and police burgh, 11 miles south of Wigtown, an ancient ecclesiastical centre, was one of the most celebrated places of pilgrimage in the country.

K

It owes its business prosperity to its rich agricultural surroundings. Its name is a corruption of the Old English word *Hwitærn*, " White House," *i.e.* St Ninian's *Candida Casa*. (pp. 79, 99, 101, 125, 128, 134.)

Ancient Sculptured Stones, Whithorn

Wigtown (1369), a royal burgh and seaport on the west side of Wigtown Bay, was one of the chief stations of the Norsemen from the eighth to the eleventh century. Its commercial importance arises from its position as centre of an agricultural district. (pp. 2, 5, 80, 85, 86, 93, 115, 120, 125, 128.)

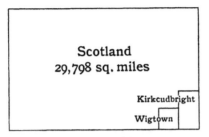

Fig. 1. Areas of Kirkcudbright (900 square miles) and Wigtown (487 square miles) compared with that of Scotland

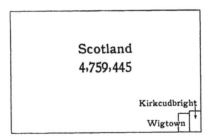

Fig. 2. Population of Kirkcudbright (38,363) and of Wigtown (31,990) compared with that of Scotland in 1911

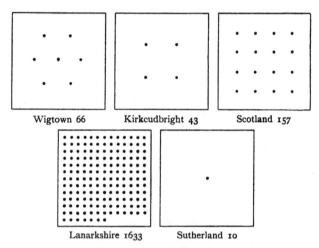

Fig. 3. Comparative density of Population to the square
mile in 1911
(Each dot represents 10 persons)

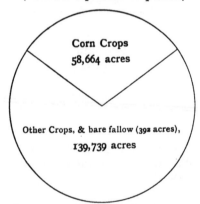

Fig. 4. Proportionate area under Corn Crops compared with that of other cultivated land in
Kirkcudbright and Wigtown in 1916

Fig. 5. Proportionate areas of Chief Cereals in
Kirkcudbright and Wigtown in 1916

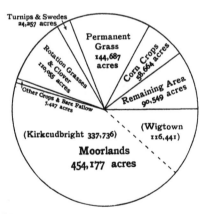

Fig. 6. Proportionate areas of land in
Kirkcudbright and Wigtown in 1916

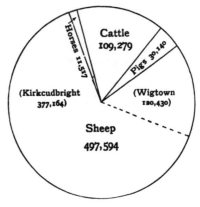

Fig. 7. Proportionate numbers of Live Stock
in Kirkcudbright and Wigtown in 1916